ECOSYSTEM
FUNCTION
& HUMAN
ACTIVITIES

ECOSYSTEM FUNCTION & HUMAN ACTIVITIES

RECONCILING ECONOMICS AND ECOLOGY

EDITED BY

R. DAVID SIMPSON
FELLOW, RESOURCES FOR THE FUTURE
WASHINGTON, D.C.

NORMAN L. CHRISTENSEN, JR
.DEAN, NICHOLAS SCHOOL OF THE ENVIRONMENT
DUKE UNIVERSITY
DURHAM, NORTH CAROLINA

CHAPMAN & HALL

I**T**P International Thomson Publishing

New York • Albany • Bonn • Boston • Cincinnati • Detroit • London • Madrid • Melbourne
Mexico City • Pacific Grove • Paris • San Francisco • Singapore • Tokyo • Toronto • Washington

Cover design: Trudi Gershenov
Cover photos: © 1997 PhotoDisc, Inc.

Copyright © 1997 by Chapman & Hall

Printed in the United States of America

For more information, contact:

Chapman & Hall
115 Fifth Avenue
New York, NY 10003

Chapman & Hall
2-6 Boundary Row
London SE1 8HN
England

Thomas Nelson Australia
102 Dodds Street
South Melbourne, 3205
Victoria, Australia

Chapman & Hall GmbH
Postfach 100 263
D-69442 Weinheim
Germany

International Thomson Editores
Campos Eliseos 385, Piso 7
Col. Polanco
11560 Mexico D. F.
Mexico

International Thomson Publishing - Japan
Hirakawacho-cho Kyowa Building, 3F
1-2-1 Hirakawacho-cho
Chiyoda-ku, 102 Tokyo
Japan

International Thomson Publishing Asia
221 Henderson Road #05-10
Henderson Building
Singapore 0315

1 2 3 4 5 6 7 8 9 10 XXX 01 00 99 98 97

Library of Congress Cataloging-in-Publication Data

Ecosystem function and human activities : reconciling economics and ecology / edited by R. David Simpson and Norman L. Christensen, Jr.
    p. cm.
Includes bibliographical references and index.
ISBN 0-412-09671-4 (alk. paper)
    1. Environmental economics —Chesapeake Bay Region (Md. and Va.)— Congresses. 2. Estuarian ecology —Chesapeake Bay Region (Md. and Va.)—Congresses. I. Simpson, Ralph David. II. Christensen, Norman L., 1946- .
HC107.A123E28   1996
333. 7'09163' 47—dc20
                          96-28143
                          CIP

British Library Cataloguing in Publication Data available

To order this or any other Chapman & Hall book, please contact **International Thomson Publishing, 7625 Empire Drive, Florence, KY 41042.** Phone: (606) 525-6600 or 1-800-842-3636. Fax: (606) 525-7778. e-mail: order@chaphall.com.

For a complete listing of Chapman & Hall's titles, send your request to **Chapman & Hall, Dept. BC,**

Contents

Preface

R. David Simpson

Norman L. Christensen, Jr.

Human Activity and Ecosystem Function: Reconciling Economics and Ecology

Recognizing the need to improve social decision making on tradeoffs between economic growth and ecological health, the Renewable Natural Resources Foundation convened a workshop in October 1995 on "Human Activity and Ecosystem Function: Reconciling Economics and Ecology." While the subtitle perhaps reflected unrealistic expectations, the presentations and discussions at the workshop were a preliminary step toward that reconciliation: bringing together ecologists, economists, other natural and social scientists, and policy makers to lay out the issues, articulate their needs and perspectives, and identify common ground for further work. This volume contains the papers presented and reports generated from the workshop.

We emphasize ecology and economics in this discussion. We could argue that organizing our inquiry around these disciplines is only natural. Ecology is the study of behavior of organisms within complex systems composed of a myriad of other organisms and their physical environments. Increasingly, this discipline has focused on how interactions among biological and physical components influence the overall functioning of ecosystems. These components are increasingly being determined by human activities. Economics is the study of how we decide which of our needs and wants we choose to satisfy given our limited resources. Thus, environmental and conservation policy decisions involve tradeoffs between the extraction or use of goods and services provided by ecosystems on one hand, and the opportunity costs imposed by conservation of ecosystem elements not immediately or directly useful to humans on the other.

We should not suppose that a convenient way of characterizing a problem constitutes a complete set of tools for analyzing it, however. Both economics and ecology must rely heavily on inputs from the other natural and social sciences. Economic analysis might be described as an inquiry at the nexus of the psychology of preferences and the technical possibilities (and, by this, we include the biological and ecological possibilities) of production. With regard to what we have called the psychology of preferences, economists typically confine their attention to what people want, rather than why people want what they want. Understanding these "why" questions—understanding that is crucial in designing environmental and conservation policies—is the province of psychology, sociology, anthropology, and even of ethics, esthetics, and theology. Also, the translation of economic analysis into practical policy requires an understanding of the political system, the province of political science. Of course, understanding the technology of production requires recognition of the physical, chemical, biological, and ecological (and, many would add, sociological and psychological) constraints on production possibilities.

Because of its interdisciplinary character, ecology has been called a metadiscipline. It relies heavily on the fields of physiology and genetics to provide an understanding of the basic workings of organisms. Physical sciences such as chemistry and geology provide the constructs and tools necessary to characterize the non-living portions of the environment and their interactions with the biota. The social sciences have contributed quantitative methods for sampling populations and characterizing variation to ecologists. In recent years, ecologists have come to understand the ubiquity and importance of historical patterns of human impacts on ecosystems and increased their collaborations with historians and historical geographers.

The organizers of the workshop, then, made a concerted effort to assemble a diverse range of presenters and participants. We hope this volume incorporates the diversity of insights represented at the workshop. Equally importantly, we hope this volume will prove valuable to the entire range of natural and social scientists, resource managers, policy makers, and concerned citizens whose efforts will be necessary to resolve the problems it addresses.

While considerable emphasis has been placed on the great geographical and temporal extent of modern environmental issues, one might paraphrase from the political aphorism and argue that all ecology—and economics—is local. This is not to contradict what we have said above, but rather to point out that the sources, and ultimately the resolution, of environmental problems lie at the level of individual people and businesses. The overworked slogan "think globally, act locally" captures the notion. Losses from environmental degradation may be felt at great distances from their sources. The solution to such problems requires that performers of environmentally damaging actions be cajoled or compelled to account for the consequences (in economic parlance, the "externalities") arising from their conduct.

Recognizing the need to consider the local sources of broader environmental problems, we focused a substantial part of the workshop on a particular case study. The Chesapeake Bay region was chosen for a number of reasons. First, it is beset by a number of problems, including overexploitation of marine resources and runoff of both toxins and fertilizers from surrounding lands. These factors have led to algal blooms, oxygen depletion, significant changes in food webs, and, as a consequence of all these factors, declines in stocks and harvest of fisheries.

The second reason the Chesapeake region makes a good case study is that it provides examples not just of ecological damage, but of social responses as well. The Chesapeake Bay watershed extends into parts of six states and the District of Columbia. Within these jurisdictions, there are scores of counties, hundreds of municipalities, thousands of businesses, and millions of households. It makes an ideal case for the study of the challenges of coordinating policy actions across jurisdictions and of the regulation of business and household environmental performance.

A third reason for examining the Chesapeake region is that extensive historical evidence is available. Native Americans had inhabited the Chesapeake watershed for thousands of years before John Smith founded the first permanent European settlement at Jameston in the early seventeenth century. One difficulty in inferring the consequences of environmental damage is that effects are not often apparent until years or decades after causes are initiated. In the Chesapeake, scientists can examine sedimentary evidence, and historians can consult written records of almost four centuries.

Finally, in recent years the Chesapeake region has attracted the attention of many leading figures in the natural and social sciences. There is now a considerable body of both natural and social science expertise and policy-making experience from which to draw. In these case studies, our objective is to draw inferences that will be useful to others doing work in other settings. We have chosen to use the Chesapeake as a case study not to identify factors peculiar to that region, but rather, to illustrate general principles that will be of interest to people working in other areas.

An Overview of Contributions

In their overview of ecosystem function, Christensen and Franklin suggest that ecosystems be defined (after Likens 1992) as "spatially explicit units of the Earth that include all of the organisms, along with all components of the abiotic environment within its boundaries." Operationally, ecologists define the boundaries of an ecosystem so as to measure or manipulate most easily the properties or processes of interest. Thus, watersheds and lakes provide ideal study ecosystems for ecologists interested in processes driven by the movement of water. Regardless of the boundaries or the spatial scale, ecosystems are always open with respect to the inputs and outputs of energy and matter.

A variety of ecosystem processes is critical to the sustained functioning of ecosystems, including conversion of solar energy to chemical energy by plants, the cycling of elements such as carbon, nitrogen, and oxygen, and the flow and transfer of water. Christensen and Franklin call attention to four key issues related to the functioning of ecosystems. First, ecological processes operate over a range of spatial and temporal scales; scales of time or space appropriate for the study or management of one process may not be the same for other processes. Second, ecosystem functioning depends on ecosystem structure, complexity, and diversity; such complexity underlies the complexity of ecosystem function, imparts both resistance to and resilience from disturbances, provides long-term capacity for adaptation, and is a sensitive indicator of environmental change. Third, ecosystems are dynamic in time and space; change is the normal course of events for ecosystems. Natural or human-caused disturbance on landscapes creates a patchwork mosaic, and the resulting changes initiated within each patch are influenced by the pattern and behavior of surrounding patches. This landscape variation heavily influences ecosystem functioning at large spatial scales. Fourth, uncertainty and surprise are inevitable. There is much we do not understand about ecosystems. Some of that ignorance will yield to increased knowledge, but the complexity and interactions of nonlinear processes promise that certain elements of ecosystem function will always be difficult to predict and that surprises in ecosystem behavior are inevitable.

Sustained management of ecosystem functioning requires attention to seven elements.

1. *Goals:* Clear, operational goals must be set that are based on the capacity of the system to provide goods and services rather than perceived or real needs for goods and services. These goals should be articulated in terms of the critical ecosystem processes and structures.
2. *Models:* Management should be informed by sound ecological models, i.e., our best understanding of how the world works. That understanding is always provisional and subject to change.
3. *Complexity:* Much extractive management seeks to simplify systems in order to concentrate productivity in those elements of greatest interest to humans. This strategy may result in high short-term yields, but carries high long-term risk.
4. *Change:* Sustainability does not imply the status quo. Change in ecosystems is necessary and inevitable.
5. *Context and scale:* Our ignorance of the importance of processes operating at different spatial and temporal scales has allowed society to define the boundaries of management jurisdictions with little or no reference to the scale of such processes.
6. *Humans as ecosystem components:* Human effects on ecosystems are ubiquitous. Humans must be recognized as integral ecosystem components directly involved in defining and achieving sustainable management goals.

7. *Adaptability and accountability:* Our knowledge base is necessarily limited and subject to change. Managers must be able to adapt to new information and understanding. Surprises cannot be eliminated; rather, management strategies should acknowledge that surprises are inevitable.

Michael Toman addresses "Ecosystem Valuation: An Overview of Issues and Uncertainties" in Chapter 2. As Toman notes, the very concept of "valuing" ecosystems may seem foreign to those who see these values as self-evident. Yet society faces choices between the preservation of ecosystems and other desires and needs. These choices must be made on the basis of some criterion. The criterion offered by the economic paradigm is "value on the margin," the rate at which we are willing to trade off a little bit of one thing to get a little bit of something else. Toman notes that the application of economic valuation is subject to both practical and conceptual challenges, however. With respect to the former, the very fact that ecosystem services tend not to be "marketed goods" implies that there is a paucity of economic data: goods that are not bought and sold have no price tags attached, so we can, at best, only determine their values to people on the basis of indirect evidence. Studies using indirect methods, such as travel cost and hedonic price techniques, often suffer from the limitations of the data on which they must rely. Survey studies—the "contingent valuation method"—generate data directly, but also generate controversy about whether respondents understand questions fully, take the exercise seriously, and offer honest answers.

On the conceptual level, some critics of economic valuation methods have faulted the emphasis placed on individual preferences as indicators of social, or even individual, welfare. Toman finds this criticism less than compelling; he points out that economic analysis is based on the axioms of preference, and axioms do not, by definition, admit direct proof or refutation. Their plausibility can only be inferred from their descriptive and prescriptive power, and on these points Toman suggests that critics have been too harsh. A more telling criticism may be that, while preferences are not irrelevant to making social choices, they may be inadequate. If ecological problems are too complex and ecological consequences too permanent for people to fathom, alternative decision processes—"constitutional rules," as opposed to "market allocations"—may be in order. Toman finds it doubtful, however, that wholly alternative measures—as, for example, embodied energy—will provide useful criteria for social decision making.

Ecological risk assessment might be considered a hybrid of ecological and economic approaches. It also brings the weight of statistical decision theory to bear on problems of ecosystem valuation. As Steven Bartell explains in his chapter on "Ecological Risk Assessment and Ecosystem Valuation," an assessment of ecological risk may be thought of in terms of a combination of three numbers: the definition of an event (e.g., a tanker runs aground), the probability with which this event happens, and the ecological and economic consequences of such an event (e.g., waterfowl are killed, fisheries are damaged). Thinking about such an "ordered triplet" is a useful way to consider the needs for information in performing ecological risk assessment.

Ecological risk assessment is a relatively young science, and progress is being made along a number of fronts in improving its applicability and performance. Bartell identifies one crucial issue as the definition of a reference environment. At the most basic level, this means the choice of a model of nature. In the past it was often supposed that ecosystems were in perpetual equilibrium: that human impacts could not overcome the regenerative power of nature, and the system would simply continue to replicate itself. This view has been displaced in many instances by a dynamic model, in which an ecosystem is not presumed to remain forever the same, but rather to oscillate in response to shocks. Management efforts might then be directed toward maintaining ecosystems within their normal ranges of variation. A third view is that ecosystems evolve from one state to another in response to both human-induced and other stresses. The goal of management under this view is not to keep things as they are, or to constrain ecosystems within historical bounds, but rather to retain the capability of the ecosystem to adapt. An important goal of management, then, would be to maintain the diversity (and particularly the genetic diversity) of organisms within a system.

Inasmuch as the conduct of an ecological risk assessment requires assigning values to events, ecological risk assessment overlaps into the area of economic valuation. Bartell touches on some of the same methods and concerns as Toman, but the approaches of the two authors are complementary. It is particularly interesting to think about the layers of value encountered in contemplating the incentives to preserve specific resources. Bartell notes as an example that Chesapeake Bay oysters are valuable as a commercial commodity, valuable in the provision of water filtration services that sustain other life in the system and, in the estimation of some, possessed of an intrinsic worth that must be respected. Bartell notes also that sustainable environmental management—maintaining the quality and quantity of ecological resources at present levels—may provide a workable objective for policy, suggesting agreement with Toman's observation that society might prefer the maintenance of a "safe minimum standard" to a case-by-case analysis of environmental costs and benefits.

The first of the six chapters written on issues pertaining to the Chesapeake Bay and its watershed is by Walter Boynton. Boynton begins his chapter on "Estuarine Ecosystem Issues on the Chesapeake Bay" by noting that estuarine ecosystems have an economic and social importance greatly disproportionate to their geographical extent. While comprising only about one half of one percent of total world ocean area, estuaries provide over a fifth of total fisheries production. Because cities and large urban populations are often located near estuaries, both human impacts and human reliance on estuary ecosystems are great.

The Chesapeake Bay provides a good example both of the stresses humans place on ecosystems and the steps society can take to relieve these stresses. Boynton notes that the Chesapeake was for many years a focus of both commercial and recreational activities. Growing populations, sprawling development, poor planning, and consequent sediment, nutrient, and toxic pollution have all had deleterious effects on the Bay. While these problems have only been recognized relatively recently, steps are now being take to reverse them.

Estuaries may be of several types, which differ by age, physical characteristics, and geological formation. The Chesapeake Bay is an estuary of the coastal plain type, formed by the flooding of a river valley at the end of the last glacial period. An exceptional physical characteristic of the Chesapeake Bay is the extent of its drainage basin relative to the volume of water in the Bay itself. As a result of this feature, pollutants accruing from runoff can become very concentrated.

Boynton emphasizes the importance of circulation patterns in understanding the Chesapeake Bay ecosystem. Because incoming freshwater does not completely mix with the saltwater in the Bay, water in the Bay is typically stratified despite its relatively shallow depth. The deeper water may become both depleted of oxygen and retained for relatively long periods of time.

Boynton argues that this high retention can be both a blessing and a curse: nutrients tend to stay in the Bay, supporting high biological productivity, rather than being washed to sea. Regrettably, however, pollutants are also retained and concentrated.

It can be difficult to distinguish between the effects of human-induced stresses on the Bay and the consequences of natural fluctuations. Rates of freshwater (and accompanying runoff) influx can vary substantially between seasons and between years. While scientific disagreement continues about the number and extent of human-induced ecological threats to the Bay's health, Boynton discusses three illustrative issues concerning which a scientific consensus is emerging. The first is the relationship between overfertilization, algal blooms, and oxygen depletion. Agricultural fertilizer runoff has led to increased concentration of nitrogen and phosphorus in the Chesapeake. This leads, in turn, to algal blooms, which, when they decompose, result in dissolved oxygen depletion and damage to marine animal stocks.

A second major ecological issue in the Chesapeake is a dieback of seagrasses. This may also be traced, at least in part, to overfertilization. Algal blooms block sunlight, stunting the growth of submerged aquatic vegetation. Once a decline in seagrasses begins its effects can be self-augmenting; among other things, submerged vegetation maintains water clarity by binding sediments and baffling wave turbulence. The murkier the water, the less the submerged vegetation is able to thrive.

The consequences of both overfertilization and seagrass declines are adverse for fisheries. With respect to the former, oxygen depletion is obviously harmful to marine animals. With respect to the latter, the loss of submerged vegetation deprives juvenile organisms of habitats conducive to their rapid growth. Boynton points out that the Chesapeake remains an "immense protein factory," despite changes to the Bay's ecology. However, there have been important losses in commercially and recreationally important fisheries. In addition to the factors mentioned above, Boynton concludes that overfishing may have been an important factor in some cases. In other instances, however, the real problem may be that fish whose natural ranges are great—especially those that spawn far inland—may be as vulnerable as the weakest element of the chain of their habitats. Improving water quality, or restricting fishing in the Chesapeake Bay itself, may have only limited effects in enlarging stocks of fish such as American shad. The shad is vulnerable to damming of the tributaries on which it spawns, acid precipitation in those streams, and fishing pressure in the waters adjoining other coastal states to which it migrates.

While the Chesapeake Bay ecosystem faces a number of threats, there is an emerging consensus that reducing nutrient runoff will alleviate many problems. The Chesapeake Bay Program, a state/federal partnership, is beginning to take steps to restore the Bay. While addressing other pollutants (for example, acid precipitation) and fishery restoration will be more difficult, Boynton is encouraged by the support for improvements in the Bay environment at local, state, and federal levels.

In their chapter, "Riparian and Terrestrial Issues in the Chesapeake: A Landscape Management Perspective," Curtis Bohlen and Rupert Friday show that human activities in the Chesapeake watershed—and particularly, events following European settlement in the seventeenth century—have worked important changes on both terrestrial and estuarine ecosystems. As population pressures continue to grow, Bohlen and Friday argue, landscape and watershed management must improve if continuing ecological degradation is to be avoided.

Bohlen and Friday emphasize a hierarchical approach to the study of landscape dynamics. Larger landscapes are composed of smaller landscapes, smaller landscapes are composed of ecosystems, and ecosystems are themselves composed of particular areas or habitats. Ecological functions at these different hierarchical levels may be characterized by different frequencies and extent; season-to-season variation in agricultural usage may dramatically affect the ecological status of any given piece of farmland, for example, but have little effect on an overall landscape composed of agricultural lands and nature reserves. An important implication of the hierarchical nature of landscapes is that the scale at which human management occurs may have little relationship to the scale of biological activities. One cannot, for example, simply decide to manage a small area of land for the production of migratory fish, when such management would require the coordination of numerous landowners (and, in fact, of sovereign nations in regulating fishing fleets on the high seas).

Bohlen and Friday illustrate their points by focusing on two contrasting landscapes in the Chesapeake watershed. Anne Arundel County, Maryland, is becoming more suburban. The Nanticoke River watershed in Maryland and Delaware remains largely agricultural. The expansion of residential construction in Anne Arundel County has resulted in greater releases of nutrients, sediments, and other pollutants. The suburbanization of the county is also causing physical and hydrological effects which have their own ecological consequences. The increase in impervious surfaces—roads, parking lots, and roofs—prevents water from filtering into soil. Less precipitation makes its way into groundwater, and more is swept rapidly into ditches, storm drains, and local streams. The "flashy" streams created by this rapid funneling of water are highly erosive, with consequences both for the deposit of sediment and pollutants downstream and the maintenance of healthy ecosystems along their banks.

In the Nanticoke watershed, in contrast to Anne Arundel County, most land remains in agriculture, forest, or wetlands. The agricultural areas tend to be maintained in what are, in ecological terms, "early successional states." Such ecosystems can be highly productive, but they also tend to be very "leaky"; nutrients and sediments are easily carried away. Such conditions lead to the runoff, overfertilization, and consequent problems described by Boynton. The agricultural regions of the Nanticoke watershed also share a feature in common with the suburbs of Anne Arundel County: the construction of drainage ditches to channel storm waters away from agriculture fields also leads to the formation of "flashy" streams and their ecological consequences.

Bohlen and Friday suggest that landscapes can be maintained by natural or cultural processes. The former are self-explanatory; these natural processes are simply the means by which ecosystems tend to replicate themselves. The latter are defined by human activity such as efforts to perpetuate the replication of landscapes which would revert to different states if not subject to human intervention. Bohlen and Friday suggest that efforts to improve the health of the Chesapeake ecosystem might involve either "environmental enhancement"—efforts to increase the output of particularly desired environmental outputs (e.g., fisheries), or "environmental restoration"—efforts to reestablish conditions under which a more "natural" (and desirable) state can exist.

In the suburban setting of Anne Arundel County, there may be little scope for environmental restoration; the costs of residential relocation may be simply too high. There are opportunities for environmental enhancement, however. For example, storm drainage systems can, be made less "flashy" by instituting measures to slow water runoff. In the Nanticoke watershed, on the other hand, restoration efforts may be undertaken. Reversion to natural landscapes in which water is allowed to settle rather than being forced to run off could ameliorate environmental damages. Of course, drainage systems were first installed precisely because farmers wanted standing water drained away. A combination of natural and cultural management—draining water into areas in which it may be allowed to settle, for example—suggests a workable solution.

There is often an implicit assumption made in discussions of environmental issues that the best of all ecological states is the "natural" one. This begs the questions of what constitutes a "natural state," and to what extent human activities are consistent with the existence of a "natural state." Grace Brush's chapter on "History and Impact of Human Activities on Chesapeake Bay" sheds considerable light on these issues. The Chesapeake Bay is geographically a relatively recent phenomenon, having been formed with a rise in sea level that followed the last ice age. The Bay has experienced a number of changes in response to climatic fluctuations over the past 10,000 years, but Brush argues that the changes in the Chesapeake ecosystem brought about by human, and specifically recent activities differ dramatically from those of earlier eras. Organisms inhabiting the Chesapeake ecosystem for most of its existence adapted to the fluctuations of the natural environment: the pace of human-induced change has been so rapid, however, that many organisms have been unable to keep pace.

Brush bases her analysis on extensive stratigraphic sampling of the Chesapeake Bay. A variety of forest types have succeeded one another in the 12,000 year history of the Bay; the region was almost entirely forested before European settlement. Land use changed rapidly following the arrival of the Europeans, with 80 percent of the land completely deforested in less than 150 years. Sedimentation rates responded accordingly, with deposition two to three times higher after land clearing began. Fertilization rates also increased, especially in response to sewage discharges. Recent years have seen large declines in submerged vegetation. Given the evidence of historically unprecedented changes in diversity and composition of species in the Chesapeake Bay, Brush suggests that restoration of the Bay ecosystem will require at least partial restoration of the terrestrial ecosystem in its watershed. This means restoring the forests and wet-lands that buffered the estuary in the millennia before European settlement.

Just as Brush documents the reciprocal interaction between human land use and ecological change, Jacqueline Geoghegan and Nancy Bockstael suggest a reciprocal connection between ecological and economic change. They report on the initial phas-es of an ambitious project to intregrate ecological and economic analysis in "Human Behavior and Ecosystem Valuation: An Ap-plication to the Patuxent Watershed of the Chesapeake Bay." Economists often model the value of land—and, by extension, the benefits of owning land—on the basis of the amenities it af-fords its owner. These amenities may include the ecological at-tributes of the area. Ecologists have devoted considerable atten-tion to the effects of human activities on ecological functioning. As Geoghegan and Bockstael note, however, there has been little consideration of the interactions between ecological and eco-nomic phenomena. Yet the connections are clear: a change in the ecological attributes of a community (whether human or other biological) affects the value of properties in that community. This change in property values may induce more or less economic de-velopment, which may in turn induce more ecological changes, and so on.

Geoghegan and Bockstael have been gathering data and con-ducting analyses of economic and ecological interactions in the Patuxent watershed. Preliminary results indicate a dependence of property values and land use on the amenity values provided by the properties. As this information is incorporated with eco-logical models linking land use to the generation of environmen-tal amenities, a complete description of the interrelated econom-ic and ecological evolution of the region will emerge.

Toman notes that economic value is based on the willingness of consumers and producers to trade one good or service (whether natural or human-made) for another. Toman, Bartell, and Geoghegan and Bockstael discuss methods by which these tradeoffs might be quantified. Henry Peskin, in the chapter on "'Green' Accounting For the Chesapeake Bay," discusses the ag-gregation of such measures into meaningful national or regional income statistics.

"'Green accounting," Peskin notes, might consist of a number of exercises. These include recording pollution abatement or cleanup expenditures, developing physical indicators of environmental quality, reporting alternative measures of national economic performance, or alternatively, well-being (e.g., conventionally measured gross domestic product less pollution abatement or cleanup expenditures), or fully integrating environmental effects into the System of National Accounts that now specifies the generation of national income and product statistics. It is this last category to which Peskin devotes most attention.

It is important to be clear on what revisions in national accounting practice might accomplish, however. On one hand, economists recognize that true measures of national well-being simply cannot, and can never, be generated from observable economic data. On the other hand, existing national accounting practice is severely deficient not only with respect to treatment of the environment, but, arguably even more so, with respect to the treatment of household labor activities and investment in education and training. Moreover, Peskin emphasizes that we need to be realistic about the uses of improved national accounts, even given their limitations. He suggests that national account information can be useful for both "scorekeeping" and "management." The former is an ex post facto tallying of performance; an intellectually challenging activity, but one with limited potential for improving performance. "Management" refers to the planning and implementation of actions to improve the functioning of the economy. The benefits of improved national accounting might not lie so much in getting better final measures of economic performance as in improving estimates of value. Better public and private management practices could result from these improved valuations.

Peskin reports the results of a study on the environmentally corrected income and product accounts of "Chesapeaka," a region composed of counties bordering the Chesapeake Bay and the estuaries of its tributaries. Peskin finds, in this region as in others he and his colleagues have studied, that the corrections made to conventional accounts to correct for environmental losses (or gains, in some cases) are small relative to overall economic measures. Surprisingly, however, Peskin finds that values arising from ecological services (for example, beach use, wildlife use, and boating) are several times greater than are measured losses from pollution.

In his conclusion, Peskin argues that "green accounting" exercises may be both more easily and credibly done, and have greater impact on policy, if incorporated into and performed by the government agencies now charged with monitoring economic performance. These agencies typically have better data (in the form of census reports, etc.) than do the nongovernmental organizations now doing most green accounting studies. In addition, issuance by government agencies would lend an imprimatur of impartiality and validity.

The final paper on the Chesapeake ecosystem is written by Timothy Hennessey. The scientific and economic perspectives advanced in the other papers on the Chesapeake region ask how policy makers weigh competing claims and establish rules for preservation or exploitation of resources. Hennessey describes the history and function of the Chesapeake Bay Program, the governance system established to manage efforts to protect and restore Chesapeake Bay. Hennessey describes three ecological attributes that make the management of ecosystems complex. First, ecosystems are highly interdependent; second, they are irreducible in the sense that management of constituent parts alone cannot insure the successful management of the whole; and third, they are subject to great temporal and spatial variability.

The complexity of ecosystems supports an adaptive approach to management. Hennessey argues that interventions must be regarded as experiments, intended certainly to improve conditions, but also to learn how ecosystems respond to management interventions. Hennessey also argues that management must reconcile and be informed by bounded conflict. Inasmuch as ecosystems provide public goods to which competing would-be users lay claim, policy makers strive not only to resolve such claims, but to learn from them.

These principles of adaptive management and bounded conflict are illustrated in the evolution of administrative mechanisms for resolving environmental conflicts in the Chesapeake watershed and regulating ongoing activity there. Hennessey identifies three stages through which policy makers passed before arriving at the present system. The first stage he characterizes as agenda setting. In this stage, scientific input was gathered and evaluated to establish priorities for policy actions. In the second stage, governance structures were established and management initiatives begun. Finally, programs were implemented and evaluated. While it is still too early to tell with respect to many resources and problems, limited evidence suggests that progress is being made, for example in concentrations of fertilizers, health of aquatic vegetation, and recovery of fisheries.

Perhaps the most important of the points Hennessey raises is that the Chesapeake Bay Program represents a link in a "diffusion of innovations." Lessons learned in the Great Lakes Program were incorporated into the Chesapeake Program, and lessons from the Chesapeake Bay Program have been incorporated in the National Estuary Program. Adaptive management may evolve in response not only to experience in the program at hand, but also to experience in other areas.

The final three chapters in this volume were written following the conference. The authors of these chapters led discussions among interdisciplinary groups on preselected topics. In his summary, Dennis Whigham contrasts ecosystem functions or processes and ecosystem values. Although challenges and issues often present themselves in terms of specific ecosystem elements such as water quality or decline of a particular species, we must focus on entire ecosystems if we are to meet these challenges successfully. Whigham argues convincingly that we must carefully distinguish between "what ecosystems do," i.e. functions, from the goods and services that human societies receive from ecosystems, i.e. values. In Whigham's chapter, the word "value" takes on an aspect different from its use in the economic lexicon. "Functions are the processes that are necessary for the self-maintenance of ecosystems, whereas values are the rules that determine what people consider important." Thus, a suite of hydrologic processes (functions) underlies the delivery of high-quality water (an ecosystem value).

Whigham argues that an understanding of the dependency of ecosystem values that include natural resources, water quality, flood control, and environmental health on ecosystem functions is critical if the feedbacks between human actions and ecosystem functioning are to be "constructive" rather than "destructive." He points out that, too often, the debate over the management of ecosystems focuses on delineation of boundaries (e.g., defining a wetland) and not on the functional attributes of the ecosystem that must be preserved. As Christensen and Franklin note, nature rarely provides us with distinct boundaries, and the geographic and temporal domains of various ecosystem functions vary considerably. Functional assessments of ecosystems such as wetlands have proven to be a challenge, but progress is being made.

Myrick Freeman's chapter, "On Valuing the Services and Functions of Ecosystems," begins with the telling observation that economists and ecologists (and others) often begin with different definitions of "value." To economists, the term generally connotes a notion of relative worth, a sense of "how much" something is worth. To others, the term often means intrinsic worth, a notion that does not admit tradeoffs. This distinction carries over into the nature of things to be valued. Ecologists may value the functions of an ecosystem quite distinct from any services arising from them. To an economist, it is the latter that are more likely to comprise the sources of value. The main thrust of Freeman's chapter is, then, toward the identification of the functions of ecosystems that give rise to flows of services of social importance. In a sense, Freeman and Whigham meet in the middle in their respective explorations of function and value. Freeman considers how economists quantify the intermediate "values" of "functions" giving rise to final goods and services; Whigham considers how ecologists define the final "values" arising from the intermediate "functions" of ecosystems.

Freeman classifies these valuable service flows into four types. First, ecosystems provide material inputs into production of other goods and services; second, they provide necessities for the support of human and other life; third, they provide for a variety of "non-consumptive" uses, such as scenic enjoyment; and finally, they provide a mechanism for the transport, storage, and eventual decomposition of waste products. In some cases—primarily those involving the first set of value sources—economists can estimate values directly from observed prices. In other cases, values must be inferred from the effects of the presence or absence of unmarketed ecosystem services consumed in conjunction with other goods. As the connection between ecosystem services and goods bought and sold in markets becomes more tenuous, the ability to assign economic value also becomes more tenuous. Freeman concludes, however, with a number of examples in which these connections might be drawn more clearly and economic analysis might be more usefully applied to policy making with respect to ecosystems.

In the concluding paper in this volume, "Ecology and Public Policy," Michael Orbach argues that all environmental policy is necessarily social policy. Input from the natural and social sciences may be weighed in making decisions about what to preserve and what to exploit. In the final analysis, however, decisions are made on the basis of social interactions and dynamics. In analyzing these factors, Orbach suggests a meaning of the word "value" different than either of those put forward by Freeman. To Orbach, the term describes a cultural rule or standard underlying decision making. In this sense of how we think, or how we behave, "values" may give rise to strict policies for, for example, the preservation of endangered species, rather than flexible tradeoffs between one desirable end and another (although Orbach's discussion of certain rigid rules might be compared to Toman's mention of the need to maintain certain safe minimum standards).

It is the expression, implementation, and, on some occasions, Orbach argues, formation or modification of values that the social decision-making process accomplishes. This process is political, an adjective which Orbach uses without pejorative intent. The political process involves a weighing of the interests of sometimes opposed constituencies. It also involves a weighing of scientific input. Scientific data may provide a factual basis on which to make decisions, but, in a telling turn of phrase, Orbach notes that society does not solve problems so much as resolve them. There remains a social decision—a value judgement in the sense in which Orbach uses "value"—as to what outcomes are preferred.

Orbach suggests three principles upon which the inclusion of science in environmental policy-making should be based. First, a total ecosystems perspective should be embraced; this implies integration across physical and human ecological boundaries, economic sectors, scientific disciplines, and governance structures. Second, the democratic nature of governance systems should be both recognized and respected. Scientists cannot arrogate to themselves decision-making authority. While a scientist may, and arguably sometimes should, choose to function as an advocate, she must distinguish, and make clear to others, whether her chosen role is as impartial source of data, administrator, or advocate. Third, scientific participants in social decision making must recognize the need for translation of scientific input into forms accessible to other constituencies.

Several lessons emerge from this collection of papers. First, it is clear that both environmental economics and ecosystem science are young disciplines within which basic concepts may be formed, but methodologies and tools are rapidly evolving. Second, issues of complexity and uncertainty loom large in both fields. Some uncertainty will yield to new data and refined understanding, but much is imbedded in the complex and often nonlinear relationships among ecosystem constituents. Third, while no "silver bullet" emerged to unify these disciplines, key areas of possible intersection such as the use of cost-benefit analyses and measures of ecosystem-level risk show promise for future collaboration between economists and ecologists. Finally, there can be no doubt that such collaborations present special challenges and compelling opportunities for the sustainable management of ecosystems in which humans will play an ever increasing role.

References

Likens, G. 1992. *An Ecosystem Approach: Its Use and Abuse.* Excellence in Ecology, Book 3. Ecology Institute, Oldendorf/Luhe, Germany.

Acknowledgments

The Renewable Natural Resources Foundation workshop on Human Activity and Ecosystem Function was made possible by funding from The Forest Service of the U.S. Department of Agriculture, the Natural Resources Conservation Service of USDA, the U.S. Bureau of Land Management, the U.S. National Biological Service, and the U.S. National Oceanic and Atmospheric Administration.

A number of individuals made valuable contributions to the planning of the workshop. We would like to thank particularly Robert Cannell, Susan Essig, Susan Everett, Glenn Flittner, Hardin Glascock, Clare Hendee, Donna Loop, Lawrence Pettinger, Priscilla Reining, Allen Rutberg, Marc Safley, Gerald Seinwill, and Keith Wadman.

The Renewable Natural Resources Foundation provided financial support, logistic and substantive assistance, and much appreciated guidance and patience. We are grateful to Norah Davis, Ralph Didriksen, and especially Robert Day of RNRF.

In addition to the authors of the papers, several other people helped by chairing, recording, and participating in discussion sessions. We are grateful to Hannah Cortner, Emery Roe, John Hendee, and Paul Portney for chairing sessions. Comments and suggestions in discussion sessions were contributed by Paul Blankenhorn, Nadine Cavender, Betsy Cody, Richard Duesterhaus, Susan Essig, John Fedkiw, Allen Fitzsimmons, Peter Garrett, Elizabeth Gillelan, Clare Hendee, Janette Kaiser, Ruthann Knudson, Raymond Kopp, Douglas Lawrence, Gene Lessard, Maurice Mausbach, Angela Nugent, Maurice Nyquist, Curtis Richardson, Jerry Ritchie, Marc Safley, Michael Scott, Rebecca Sharitz, Willis Sibley, Max Toch, Cheryl Trott, Richard Weismiller, and Amy Wolfe. Finally, we thank Kristy Manning for able and enthusiastic assistance in the final editing process.

With support from the Renewable Natural Resources Foundation Member Organizations: American Anthropological Association, American Congress on Surveying and Mapping, American Fisheries Society, American Geophysical Union, American Meteorological Society, American Society for Photogrammetry and Remote Sensing, American Society of Agronomy, American Society of Civil Engineers, American Society for Horticultural Science, American Society of Landscape Architects, American Society of Plant Physiologists, American Water Resources Association, Association of American Geographers, The Ecological Society of America, The Humane Society of the U.S., The Nature Conservancy, Society for Range Management, Society of Wood Science and Technology, Soil and Water Conservation Society, and The Wildlife Society.

Contributors

Steven M. Bartell
SENES Oak Ridge, Inc.
Center for Risk Analysis
102 Donner Drive
Oak Ridge, TN 37830

Nancy Bockstael
Department of Agricultural and
Resource Economics
University of Maryland
College Park, MD 20742

Curtis C. Bohlen
Department of Environmental
    Studies
Bates College
Lewiston, MN 04240

Walter R. Boynton
Professor
Chesapeake Biological
    Laboratory
Center for Environmental and
    Estuarine Studies
University of Maryland System
Box 38
Solomons, MD 20688

Grace S. Brush
Department of Geography
    and Environmental
    Engineering
Johns Hopkins University
Ames 313
3400 North Charles St.
Baltimore, MD 21218

Norman L. Christensen, Jr.
Dean
Nicholas School of the
    Environment

Duke University
Durham, NC 27708

Jerry F. Franklin
College of Forest Resources
AR-10
University of Washington
Seattle, WA 98195

A. Myrick Freeman III
Department of Economics
Bowdoin College
Brunswick, ME 04011

Rupert Friday
Chesapeake Bay Foundation
Maryland Office
164 Conduit Street
Annapolis, MD 21401

Jacqueline Geoghegan
Department of Economics
Clark University
950 Main Street
Worcester, MA 01610

Timothy M. Hennessey
Director
Rhode Island Administration
    Program
Department of Political Science
and
Associate, Coastal Resources
    Center
University of Rhode Island
Kingston, RI 02881

Michael K. Orbach
Marine Laboratory
Nicholas School of the
    Environment

Duke University
Beaufort, NC 28516-9721

Henry Peskin
Edgevale Associates
1354 Black Walnut Dr.
Nellysford, VA 22958

R. David Simpson
Fellow
Resources for the Future
1616 P Street, N.W.
Washington, DC 20036

Michael A. Toman
Resources for the Future
1616 P Street, N.W.
Washington, DC 20036

Dennis F. Whigham
Smithsonian Environmental
  Research Center
Box 28
Edgewater, MD 21037

ECOSYSTEM
FUNCTION
& HUMAN
ACTIVITIES

CHAPTER ONE

Ecosystem Function and Ecosystem Management

Norman L. Christensen, Jr.
Jerry F. Franklin

Introduction

Sustained Yield and Sustainable Management

Surely the most sobering, if not daunting, lesson to emerge from millennia of natural resource use and a century of ecological study is that the laws of conservation of mass and energy apply to the ecological systems we depend upon for natural resources and to the environmental conditions that shape the quality of our lives. Ecologists refer to the processes that shape and move matter and energy in ecosystems as "functions." Thus, we may liken ecosystem functioning to the processes that comprise physiological functioning in an organism. These functions are the basis for sustained provision of "goods" and "services" upon which humans depend (Table 1–1).[1]

Current discussion over sustainable management and development may seem odd to those who recognize that the phrase "sustained yield" has been in the natural resource lexicon for over a century. However, sustained yield management strategies historically focused on the need to deliver products to mills and suppliers as opposed to management strategies focused on the functions necessary to produce those products in the first place.

Thus, sustained-yield forestry saw as its goal the maintenance of an array of forest age-classes necessary to provide wood fiber at a continuous (i.e., sustained) rate. Only recently have we begun to worry about site and landscape processes that shape the long-term productivity of forests (e.g., Society of American Foresters 1993).

At least three trends have raised our awareness of the need for sustainable management of ecosystems. First, the level of our

Table 1-1   Ecosystem Goods and Services. Healthy ecosystems perform a diverse array of functions or processes that are critical to the sustained provision of both goods and services. Here, goods refer to products or items given monetary value in the market place, whereas the services from ecosystems are often greatly valued, but rarely bought or sold.

Ecosystem "functions" include:
- Hydrologic flux and storage
- Biological productivity
- Biogeochemical cycling and storage
- Decomposition
- Maintenance of biological diversity

Ecosystem "goods" include:
- Food
- Construction materials
- Medicinal plants
- Wild genes for domestic plants and animals
- Tourism and recreation

Ecosystem "services" include:
- Maintaining hydrological cycles
- Regulating climate
- Cleansing water and air
- Maintaining the gaseous composition of the atmosphere
- Pollinating crops and other important plants
- Generating and maintaining soils
- Storing and cycling essential nutrients
- Absorbing and detoxifying pollutants
- Providing beauty, inspiration, and research
  (Modified from Ehrlich and Ehrlich 1991, Lubchenco et al. 1993, and Richardson 1994)

ignorant bliss has decreased as our understanding of the behavior of ecosystems and their finite ability to respond to human demands has increased. Second, growth in human populations, coupled with the fact that there are few if any new frontiers for expanding our management, makes the need for sustainable management of our ecosystems truly compelling. Third, the demands on most ecosystems have become much more complex, and such demands often compete with one another. For example, sustainable forestry is not defined simply in terms of the amount of wood that can be harvested without depreciation of a site's capacity to produce wood, but also in terms of the impact of wood harvest on wildlife, water resources, and scenic beauty of landscapes. The demands for these latter commodities have increased dramatically with urban sprawl and population growth.

Ecology and Ecosystem Management

Ecologists divide themselves into subdisciplines in a roughly higher hierarchical fashion. Physiological ecologists focus on the behavior and response of individual organisms to variations in such factors as temperature, water, or the availability of other necessary resources. Population ecologists concern themselves with the patterns of growth or genetic diversity that exist among congregations of individuals that potentially interbreed (i.e., species). Interactions among species such as competition, predation, and parasitism are the domain of community ecologists.

The study of ecosystems represents the next step in this hierarchy and acknowledges the fact that the processing of matter and energy in any particular space depends on the interactions between living organisms and their physical environment. Ecosystem ecologists see the physical environment—the physical and chemical nature of soils, water, and the atmosphere—not only as affecting organisms, but also as being shaped by their activities. Over the last decade, a new level has been added to this hierarchy, landscape ecology. This subdiscipline is concerned with the importance of spatial positions and scales of ecosystems and ecosystem components relative to one another.

The phrase "ecosystem management" might suggest a favored role for ecosystem scientists in defining and informing this new field. However, sustained management of ecosystem function depends on input at all levels of this hierarchy. To understand the mechanisms that underlie behavior at one level, say, populations, we must understand the impacts of environment at lower levels of organization. say. individual organisms (physiological ecology). At the same time. we must realize that such behavior occurs in the context of a complex array of other species (communities).

Similarly, management objectives are only rarely focused on entire ecosystems. Rather, groups of humans are generally interested in specific ecosystem components such as the growth of a particular species, the quality of water. or the ability of an area to absorb and retain less waste chemicals. Nevertheless, the sustained realization of any of these goals depends on the behavior of the entire system. Furthermore. strategies to optimize particular ecosystem elements may necessarily sacrifice the behavior of others.

Defining Ecosystems

Likens (1992) defines an ecosystem as "a spatially explicit unit of the earth that includes all of the organisms. along with all components of the abiotic environment within its boundaries." This seemingly simple definition emphasizes the importance of the combination of organisms with their nonliving environment. However, it does not give us much *operational* guidance: i.e., it does not provide us with a list or means of classifying the earth's ecosystems, nor does it provide us with rules for the actual delineation of ecosystem boundaries "on the ground." Such operational rules for defining ecosystems such as wetlands have been the source of considerable debate among scholars and decision makers (Whigham, this volume).

The comparison of ecosystems to organisms (e.g.. Golley 1993) is at once apt and misleading. Functions such as the processing of matter and energy that are so much a part of the definition of an individual organism have obvious analogues in the cycling of nutrients or transfer of energy through food webs within ecosystems. The developmental process by which an individual organism grows from a single cell into a complex, highly organized array of specialized tissues has served as a metaphor for the process of ecosystem change following disturbance (Odum 1969). Such comparisons have pedagogical value, but an ecosystem is no more an organism than a map is the actual territory it represents. Just as geographers must understand the distortions created by different map projections, we should focus on the limitations of our ecological metaphors and models.

Consider how an ecosystem is not like an organism:

- Boundaries. Individual organisms usually have discrete boundaries and are easily differentiated from one another. Operationally, we can census and count them with comparative ease. Our ability to define the boundaries of ecosystems varies widely and depends specifically on the functions or values of interest.
- "Natural" classification. Four billion years of evolution and genetic differentiation have produced a diverse array of organisms which can be grouped "naturally" into species on the basis of the likely transmission of genetic information among individuals. Biologists may debate the fine points of individual classification decisions, but the overall system of organismal taxonomy is seen as "natural," i.e., resulting directly from the process of evolution. Species are components of ecosystems, but there is no genetic code that determines the exact structure of an ecosystem, nor has evolution produced categories of ecosystems that are discrete from one another in the same sense as species. In short, a "natural" classification of ecosystems is not possible.
- Development. The developmental patterns of individual organisms may vary in their degree of determinism, but the general map for such development is imbedded in a reasonably well understood DNA code.[2] No such code exists for the development and structure of an ecosystem, and the process of ecosystem development is shaped by a diverse array of factors internal and external to the ecosystem.[3]
- Structure/function relationships. Cell types of higher organisms are differentiated in tissue types which perform reasonably well understood functions critical to the survival and reproduction of an organism. It is tempting to think of the diversity of organisms in an ecosystem as being analogous to the diversity of cells or tissues in an organism; however, ecosystem functions as we shall consider them here rarely if ever have a "one for one overlay" with particular organisms.

In many ways, it is the lack of hard and fast rules for definition of ecosystems that has made the ecosystem concept useful to ecologists interested in processes such as the movement of water, cycling of carbon, or flux of energy. They recognize that, whatever scale or boundaries are selected to define an ecosystem, such systems are always "open" to the movement of energy and matter. Spatial scale and boundaries are thus selected so as to measure, monitor, or manipulate the process of interest most easily. Thus, for processes driven by the movement of water, catchments or watersheds are often ideal boundaries, while processes affected heavily by the chemistry of the atmosphere might be better studied at different scales and with different boundaries.

In similar fashion, we might define ecosystem boundaries for management purposes in ways that allow us to monitor or manage ecosystem processes most easily. Nevertheless, sustained ecosystem function at any scale depends on the input of energy and matter from other systems, and scales appropriate for managing one process may prove challenging at best and inappropriate at worst for managing others.

In asserting that ecosystems are "open" to matter and energy, in essence we are saying that the sustainability of key functions within an ecosystem depends on "subsidies" of matter and energy from surrounding ecosystems. This is true at all spatial scales, even at the level of the entire biosphere, where sustained function depends on input of solar energy at a relatively constant rate.

The intensity of human impacts associated with our management of ecosystems varies along a gradient from wildlands to urban centers, and it considerably influences the subsidies required for sustaining function and delivering goods and services (Table 1–2, Christensen et al. 1996). Intensively managed systems depend on inputs from less intensively managed systems. Thus, Baltimore is sustainable in the context of the Chesapeake watershed because it is embedded in an array of less intensively managed systems that provide essential subsidies and services.

Ecosystem Functioning in Four Lessons

Among ecologists there is a tendency, perhaps self-serving, to focus on the seeming ineffable complexity of ecosystems. "The ecosystem is a highly complex phenomenon. It is not only more complex than we think. It is more complex than we *can* think."

Table 1-2   A conceptual framework for ecosystem management goals, inputs and outputs

Category	Ecosystem Type and Human Use	Intensity and Goals of Management	Inputs	Outputs
Intensive	Urban	Intensive management to provide food and shelter for use	Heavy subsidies of energy (fossil fuels), materials (fertilizers, metal and wood) and human labor	Manufactured products Pollutants and toxins Food and water
	Intensive agriculture, human aquaculture, and suburban			
	Plantation forestry			
	Managed pasture			
Semi-natural	Managed forestry, grazing, wildlife and fisheries	Moderate management for sustained production of natural resources and for maintenance of ecosystem processes	Moderate inputs of energy, materials, and human labor	Harvested natural resources: timber livestock, fish, wildlife, water, minerals, fossil and fuels
	Forest Grassland Woodland Shrublands Lakes Streams/rivers Wetlands Estuaries Oceans			Ecosystem services
	Extraction preserves			
Natural	All kinds	Minimal management to maintain biological and habitat diversity, integrity of natural ecosystem processes and esthetic values	Minimal management to maintain near-natural conditions	Recreational uses
	Reserves and wild areas			Ecosystem services

(Egler 1974). However, while the details of many ecosystem processes and their relationships to specific ecosystem components remain poorly understood. several general "lessons" have become clear and are well substantiated by volumes of data. Here, we would like to focus on four of those lessons that are central to the sustainable management of ecosystems. Understanding of these lessons is prerequisite to assigning value (however defined) to ecosystem structure and functions.

There is No Ideal Spatial or Temporal Scale

In the best of all worlds. we would define the boundaries of natural resource jurisdictions so as to manage most easily the processes necessary to achieve our management goals. In the real world, jurisdictional boundaries have typically been established with little or no reference to ecological process. Consider how often rivers serve as boundaries between ownerships, counties, states, and countries. Rivers may have strategic significance from the standpoint of defense or simply the ease of delineation, but they bisect watersheds and, therefore, the host of processes in the ecosystem connected to the hydrologic cycle. Similarly, the boundaries of the complex set of jurisdictions that comprise the Chesapeake ecosystem match poorly the variety of processes discussed by Bohlen and Friday (this volume).

But even if we had the opportunity to reorganize jurisdictions, we would find no ideal scale or boundaries. Ideal spatial scales for the study or management of one issue, process, or element are not necessarily ideal for of others. For example, watersheds represent a useful unit for the study of water and nutrient movement driven by hydrology, but may not be ideal for studies of energy transfer through food chains in areas where animals move over large distances. In management, they may represent a useful unit for management of stream water quality but are less useful for the management of large vertebrate ungulates or carnivores.

Ecosystem Function Depends on Structure and Diversity

Ecosystem diversity is "the variety of life and its structures and processes, including the variety of living organisms and the genetic differences among them, and the communities, ecosystems and landscapes in which they occur" (Keystone Center 1991; Wilson 1992). The values humans assign to ecosystems often focus on a small subset of the organisms or functions that comprise an ecosystem. Indeed, it can be argued that a key strategy in much of human management focuses on system "simplification," the elimination of ecosystem components or processes that might diminish the availability of those ecosystem components of special interest to humans. Thus, we weed our gardens and croplands and expend considerable energy attempting to eliminate insect pests and "varmints" that compete with us for food. In addition, we have learned that human activities that alter the structure of ecosystems or vary their distribution and pattern on landscapes often alter and diminish the diversity of organisms that comprise them.

The potential and real loss of species from ecosystems has been a matter of public interest, especially since the implementation of the 1973 Endangered Species Act. The preservation of diversity is often articulated in terms of our ethical responsibility to the biota and species' rights to exist, and such preservation can be seen as having significant "opportunity costs" (e.g., "owls versus jobs") insofar as it limits our ability to exploit goods and resources.

Without diminishing ethical arguments for the conservation of the Earth's biological diversity, it should be recognized that actions that diminish the structural complexity and biological diversity of ecosystems often present real opportunity costs with regard to the sustainability of functions and, thereby, the long-term provision of goods and services. Connections between biological diversity and ecosystem sustainability include functional dependencies, resistance to and recovery from disturbance (stability), and long-term adaptability to environmental change. In addition, species diversity can be a subtle indicator of changes that may influence ecosystem health. These connections are discussed below.

1. *Functional dependencies.* However complex, the connections between biological diversity and processes within ecosystems are obvious. The flux of energy through ecosystems depends on the complexity of food webs organized by a diverse array of primary producers (plants) and secondary trophic levels (herbivores, carnivores,

decomposers, etc.). Certain groups of organisms such as nitrogen fixing microbes or sulfur oxidizing bacteria are often identified with specific functions that influence the availability of materials to other organisms. Less obvious, but equally important, is that organisms create structures that interact with and alter the physical world. Through its effects on the condensation, interception, and evapotranspiration of water, the biota regulates hydrologic processes, and geomorphic processes such as erosion and soil development often derive directly from the activities of certain organisms. Finally, complexity begets complexity; organisms provide habitat for other organisms that often fulfill critical functions.

It is tempting to ask, "How many species are needed to maintain key ecosystem functions?" or "At what rate and in what ways is function diminished by decreasing diversity?" As Christensen et al. (1996) comment, these questions assume that 1) there is some predictable overlay between species diversity and functional diversity (i.e., species functions within an ecosystem are in some sense analogous to job categories within a factory), 2) that we know all the individual organismal activities that comprise overall ecosystem function, and 3) that ecosystems do not change in ways that may influence which species are best able to carry out key functions. These assumptions are false!

Several species can play similar roles with regard to particular functions, thus the notion of "redundancy" provides a metaphor for understanding the value of diversity. However, individual species also influence a variety of functions and are not substitutable in the same sense as redundant computers or fuel valves on a space shuttle. Furthermore, functional relationships among organisms and their relative abundance can change dramatically from time to time in relationship to disturbances and environmental change (see page 14).

2.   *Resistance to and resilience from disturbance.* Biological diversity provides for both stability (resistance) for and recovery (resilience) from disturbances that disrupt important ecosystem processes (Christensen et al. 1996). The complex interactions among organisms, such as

food webs that provide alternate pathways for achieving particular flows of energy and nutrients, impart resistance to change in key functions as environments change. For example, numerous fungal species form mutually beneficial relationships with tree roots (mycorrhizae) that facilitate the cycling of key nutrients such as phosphorus. The diversity of these fungi buffers forest ecosystems against the loss of individual tree species and makes total loss of this nutrient cycling function unlikely, even in the face of enormous seasonal, annual, and longer-term climatic variations.

In many situations, diversity minimizes the risks of catastrophic change in agricultural and forest ecosystems by limiting the spread of species-specific pathogens and "pest" insects. Crop or forest monocultures may result in high-level production of specific products or resources, but they are also at much higher risk of catastrophic loss from such disturbances than their more diverse counterparts. In this context, risk analyses such as those described by Bartlett (this volume) may be useful in assigning economic value to the preservation of diversity. The importance of species diversity to the ability of ecosystems to recover ecosystem functions such as productivity following a disturbance or perturbation has been convincingly demonstrated in long-term studies of productivity responses to drought in grasslands (Tilman and Downing 1994).

3. *Long-term adaptability.* Given ever-changing environments, the capacity to adapt is central to the long-term sustainability of ecosystem function. Long-term adaptations of ecosystems to changes in climate and other environmental variables are primarily dependent upon available biological diversity. Long-term pollen profiles suggest that relatively unimportant species restricted to particular microsites during one climatic regime may become important and more widespread as the climate shifts (e.g., Brush, this volume).

4. *Indicator of environmental quality.* The diversity of ecosystems can serve as a sensitive indicator of environmental changes that can influence the production of goods and services. For example, one of the early signs of eutrophication in freshwater ecosystems is the

change in relative abundance of organisms and decrease in the number of species of planktonic algae. As
we learn more about the connections between ecosystem functions such as nutrient cycling, we can identify
specific biotic elements as indicators of the status of
those functions.

Our limited understanding of the nature of connections
among the array of ecosystem components presents daunting
challenges to managers. The responses of particular elements to
particular interventions or insults are often delayed in time and
may appear to be disconnected from the specific stimulus. Management of biological diversity is made all the more difficult by the
fact that such diversity is itself a dynamic property of ecosystems
affected by variations in spatial and temporal scale. On a relatively local scale, such as a hectare of forest or a small portion of
an estuary, species' populations may rise, fall, or even go locally
extinct as environments change. On a regional scale, such populations are generally less variable because of the connections
among habitats and the ability of species to migrate and reestablish. Clearly, human activities that influence the functional size
of populations (e.g., the size of forest patches) or their ability to
migrate (e.g., the continuity of landscapes or water bodies) may
have serious consequences for the diversity of populations.

Finally, it is important to appreciate that the values of diversity to ecosystem function are not realized merely by increasing
the number of things with no specific reference to their identity.
The redistribution of species across the globe must be viewed as
one of the most significant human impacts on ecosystems. The
negative consequences of exotic species in both natural and
managed ecosystems (e.g., Vitousek 1990) stand as stark testi-
mony to the fact that the contribution of biological diversity to
ecosystem function is not merely a function of the number of
species present.

Ecosystems Are Dynamic in Space and Time

"Ecosystem management is challenging in part because we
seek to understand and manage areas that change" (Christensen
et al. 1996). A mere 12,000 years ago, glacial ice sheets extended to the northern reaches of the Chesapeake, and the mid-Atlantic region supported expanses of boreal forests similar to

those found in Quebec and Ontario today. We are just beginning to understand the complexity and scales of change ranging from seasonal variations in rainfall, to periodic events like the *El Niño*/Southern Oscillation cycle, to long-term changes and large-scale changes, such as decades- and centuries-long drought or temperature cycles.

Disturbances, natural and human caused, are ubiquitous (Connell and Sousa 1983). Such disturbances reset succession within all or a portion of an ecosystem, leading to mosaics of successional patches of different ages within and across landscapes. Successional change may itself influence the likelihood of disturbance. For example, accumulation of woody debris in many forests and shrublands may increase their flammability, as well as the intensity of a fire when it occurs (Christensen 1994). Susceptibility to wind or ice damage changes during succession in eastern deciduous forests (White 1979).

The notion of equilibrium has inspired a long history of interest and controversy among ecologists (e.g., Egerton 1973, Bormann and Likens 1979, Connell and Sousa 1983). The concept of "homeorhesis" (see O'Neill et al. 1986), the tendency of a disturbed system to return to its predisturbance *trajectory* or rate of change rather than "homeostasis," the tendency to return to some predisturbance state, seems appropriate for ecological systems. Homeorhetic stability implies return to normal *dynamics* rather than return to an artificial "undisturbed" *state*. Ecological systems do not exhibit an undisturbed "state" that can be maintained indefinitely. Rather, they exhibit a suite of behaviors over all spatial and temporal scales, and the processes that generate these dynamics should be maintained.

If change is constant and inevitable, it does not follow that ecosystems can sustain change of any kind. Extreme fluctuation is unusual in most ecosystems and can seriously alter some ecosystem processes. Over the four billion year history of the earth's biota, the earth's environment has undergone significant change. However, it is likely that ecosystems such as the Chesapeake have never experienced change at the rate at which it is occurring today. Furthermore, many changes such as extremes of land fragmentation and certain kinds of pollution have no precedent in evolutionary history of the biota of such ecosystems. The rapidity of change as well as the novel character of many human impacts present special challenges to our ability to manage ecosystems sustainability.

Table 1-3   The "Value" of Wood

The economic value of wood is generally assumed to be that assigned to it in the marketplace for timber and fiber, and the question of when to cut a tree has been a central paradigm for the discipline of natural resource economics. Subsidies in pesticides or fertilizers, as well as investments in roads or such activities as thinnings are often including in such valuations as well as the opportunity costs for holding wood as trees, rather than cutting and selling it.

Nearly all economic analyses assume that there is no positive economic value to leaving wood, dead or alive, in a forest. Nevertheless, the structural complexity created by the variety of wood in its many stages of life, death, and decay are now known to influence a variety of ecosystem processes. Woody debris on the forest floor is often the site for unique biogeochemic reactions and may be critical for reestablishment of the forest following disturbance. For decades, forest managers "cleaned" stream channels of fallen dead trees; we now know that this material modulates hydrologic flows and creates habitat complexity necessary for a diverse array of steam organisms and the functions associated with them. Standing snags provide critical habitat for a wide variety of organisms.

Thus, while harvest of wood yields direct economic benefits, retention of both green and dead wood within an ecosystem also adds potential economic value to the ecosystem. Admittedly, we are not able at this time to quantify the values that derive from wood retention as precisely, but to the extent that they are important to the long-term sustainability of ecosystem functions, their value may be considerable.

Uncertainty and Surprise Are Inevitable

If we only knew more, we would manage our natural resources better. There is no doubt that, with increased knowledge and understanding, we could resolve key uncertainties related to management choices. Nevertheless, there are uncertainties and surprises inherent in ecosystem behavior that are unlikely to yield to more knowledge.

Unknowable responses and true surprises arise from the complex and ever changing character of ecosystems and their responses to disturbances that are in the history of the ecosystem. "Such uncertainties cannot be eliminated or reduced, but their magnitude and relative importance can be estimated. Examples of such uncertainties include ecosystem responses to unprecedented rates of climate change, carbon dioxide enrichment, or increased ultraviolet radiation. They also include such rare events as meteor impacts, earthquakes, and volcanic eruptions. Finally, uncertainties of this type may derive from cumulative effects of multiple environmental changes, such as the accumulation of insults to aquatic and marine ecosystems that have influenced populations of migratory fishes" (Christensen et al. 1996).

Sustainable use of ecosystem goods and services requires a "safety factor" to account for uncertainty. If uncertainty is reduced, utilization may be increased. But in some cases, the marginal benefit derived may be less than the cost of the research and monitoring necessary to reduce uncertainty (Walters 1986). Cost-benefit analysis of such tradeoffs must surely be a place for additional collaboration between economists and ecologists.

The Challenges of Ecosystem Management

Based on the key lessons outlined above, Christensen et al. (1996) identify eight prerequisites for the sustainable management of ecosystems (Table 1–4), as well as five challenges to the implementation of these management strategies, discussed below.

Defining Operational Goals and Objectives

While our ultimate goals for complex ecosystems may include a variety of goods and services, it is critical that *operational* management objectives be formulated in terms of the ecosystem functions and processes necessary to produce those goods and services sustainably. Such operational goals must be defined by our best understanding of the connections between ecosystem functioning and the delivery of ecosystem goods and services.

Few would consciously advocate policies or activities that would diminish the capacity of ecosystems to provide the goods and services we enjoy today to future generations. Indeed, sustained capacity and potential are legally mandated requirements for many public agencies charged with the stewardship of natural resources. However, "management that focuses on commodity resources alone, that does not acknowledge the importance of diversity and complexity, that is not aware of influences

Table 1–4 Key Elements of Ecosystem Management

1. **Setting Operational Goals:** Goals should be formulated so as to provide benchmarks for successful management policies and practices. Operational goals pay special attention to ecosystem structures and processes needed to the sustained provision of natural resource goods and services.

2. **Managing in Context and Across Scales of Space and Time:** It is in the context of unique circumstances and objectives of particular places that ecological principles are successfully applied to practical problems. The spatial and temporal context for management decisions should match the scales of ecosystem processes critical to sustainability. The connectivity of the landscape and daunting reality that actions at one location may influence ecosystem properties and processes elsewhere should be an integral part of management planning.

3. **Complexity and Diversity:** Biological diversity and structural complexity of ecosystems are critical to such ecosystem functions as primary production and nutrient cycling. Complexity and diversity also impart resistance to and resilience from disturbance, and provide the genetic resources necessary to adapt to long-term change.

4. **Variability and Change:** Forest ecosystems are constantly changing. Natural disturbances such as fire, windstorms, insect and pathogen epidemics, and floods are ubiquitous and, in many cases, critical to the maintenance of key ecosystem processes. Management determined to "freeze" ecosystems in a particular state has generally proven to be futile and unsustainable.

5. **Uncertainty and Surprise:** There are limits to the precision of our predictions set by the complex nature of ecosystem interactions. To be sustainable, management must include a margin of safety for uncertainty. Ecosystem management is not a strategy to eliminate uncertainty or surprise. Rather, it acknowledges its inevitability and accommodates it.

6. **Monitoring:** Given the limitations in our understanding of the behavior of forest ecosystems, managing without a program to monitor elements critical to our goals is akin to trying to navigate without a compass (Lee 1993). Monitoring critical ecological processes that produce these forest resources is essential to detecting changes.

(Continued)

Table 1–4 Key Elements of Ecosystem Management (Continued)

7. **Adaptability and Accountability:** Management should be viewed as experimental and include the means to learn from our experiments and adjust goals and practices accordingly. This concept of "management as experiment" is the basis for adaptive management.

8. **Human as Ecosystem Components:** Humans may present some of the most significant challenges to sustainability, they are also integral ecosystem components who must be part of any effort to achieve sustainable management goals. Given the growth in human populations, sustainable provision of ecosystem goods and services becomes an even more compelling goal.

of and impacts on surrounding areas, and that concerns itself with short time frames is not likely to be sustainable in the long term" (Christensen et al. 1996).

In many cases, models for connecting functions with commodities or services are reasonably well understood; more often they are not. Thus, a strong case can be made that with more information, our ability to set operational goals will improve. However, such goal setting is not simply information limited. The fact that the key ecosystem functions necessary for the sustained provision of some goods and services are different from and perhaps even detrimental to the provision of others presents an even more formidable challenge to sustained management of complex ecosystems such as the Chesapeake. For example, nutrient subsidies (i.e., fertilizer applications) necessary to the sustained production of crops may be detrimental to ecosystem processes that support coastal fisheries or provision of potable water in streams. Thus, because setting operational goals requires reconciling conflicts among segments of society that may have competing interests in the behavior of ecosystems, operational goals will likely not be set based on some overall vision for the optimal function of an ecosystem.

Reconciling Spatial Scales

If the spatial scales and borders of management jurisdictions were congruent with the behavior of processes central to sustained ecosystem function, sustainable management would be infinitely easier. Reconciling the various stakeholders' objectives and actions within the domain of an ecosystem must be central to the implementation of sustainable management strategies. "Because ecosystems exist at several geographic scales, so, too, should efforts to coordinate activities that affect them" (Congressional Research Service 1994).

Strategies aimed at realigning spatial boundaries may be useful in some cases, but are likely to be futile in most. It may be possible economically and politically to adjust the boundaries of a wilderness preserve in order to minimize conflicts to management goals. In such cases, goals are comparatively simple and societal constraints are few. Over larger areas, adjusting boundaries of ownership and jurisdiction often requires reconciliation of competing values and is likely to be economically prohibitive. Furthermore, given the variation that exists in the spatial domain of different ecological functions or processes, a perfect fit for different ecological functions and processes simultaneously is virtually impossible.

Identifying stakeholders represents a significant challenge relative to spatial scale. In some cases, stakeholders may be identified by simply matching maps of ownership with the appropriate ecosystem boundaries, such as the domain of a watershed. More often, stakeholders also include parties who have no title or legal jurisdiction, but are dependent on or have an interest in the goods and services an ecosystem provides. For example, communities dependent on clean water delivered from a watershed are obvious stakeholders in the management of that watershed. More often, the network of stakeholders is complex. People who never fish for striped bass or eat Maryland blue crabs may, nevertheless, have strong commitments to their conservation as evidenced by a willingness to support nongovernmental organizations focused on these issues. They too are undeniably stakeholders.

Reconciling Temporal Scales

Public and private management agencies are often forced to make fiscal-year decisions about resources whose behavior is better measured in centuries. Ecosystem management must deal with time scales that transcend human lifetimes and almost certainly exceed the timelines for other political, social, and economic agendas.

In the debate regarding the reality and directionality of human-caused environmental change, it is easy to forget that change (natural or human caused) is nevertheless inevitable. Climates vary at virtually every time scale, and processes of natural disturbance and ecosystem recovery guarantee that ecosystems are constantly changing with regard to their structure and associated functions. This change is made all the more complex by the mixture of human land uses and successional changes that typify landscapes such as the Chesapeake watershed.

Understanding temporal change in ecosystems is critical to their sustainable management. For example, in arid and semi-arid landscapes during periods of abundant rainfall, it is tempting to expand development and agricultural activity, forgetting that periods of drought are not just likely, but inevitable.

Given sufficient time, improbable events will probably happen. To implement ecosystem management, we must develop strategies that incorporate long-term planning and commitment, while recognizing the need to make short-term decisions. Ecosystem management is not an antidote for surprise; rather, it is an approach to management that acknowledges that unlikely events do happen (Holling 1993).

Making Management Systems Adaptable and Accountable

Given the uncertainties described above, it is essential that management systems be adaptable. They must adapt to variations in environment (including the impacts and needs of humans) from location to location. They must also be adaptable to inevitable changes in those environments through time. Most important, management systems must acknowledge the provisional nature of our models and information base and adapt to new information and understanding (Holling 1978; Walters 1986). To be adaptable and accountable, management objectives and expectations must be explicitly stated in operational terms, informed by our best models and understanding of ecosystem function, and tested by carefully designed monitoring programs that provide accessible and timely feedback to managers.

Management protocols and practices should be viewed as experimental and monitoring programs should be viewed as the means to test the hypothesis that these interventions are, in fact, moving the managed system toward intended goals. To do this, we must clearly state objectives in operational terms that are relevant to sustained ecological function, develop monitoring programs focused on data relevant to those operational objectives, provide for efficient analysis and management of data, and encourage timely feedback of information from research and monitoring programs to managers.

Three elements, focus, efficiency, and commitment, are critical to the success of monitoring programs (Christensen et al. 1996). Monitoring should be focused on management expectations (operational objectives) and designed to test the success and efficacy of management practices. Efficiency refers to the need for rigorous statistical sampling designs with attention to issues of precision and bias in data gathering. Because management situations often offer limited opportunities for replication or are often biased by patterns of ownership and accessibility, sampling designs will often be flawed. Such flaws should not be taken as an excuse to avoid monitoring, but their likely impacts on data quality and uncertainty of conclusions must be explicitly evaluated.

A long-term vision and commitment to the development and maintenance of monitoring programs is critical. Such programs necessarily add cost and can be especially difficult to maintain where personnel turn over frequently. Thus, clear identification of target objectives for monitoring is important; "shotgun" approaches may miss key variables while incurring the unnecessary cost of gathering irrelevant data.

The adaptive management loop will be closed only if there is timely feedback of results of monitoring information, as well as new insights from basic research programs, to managers. Such feedback will require institutional change in public agencies charged with natural resource stewardship, and may even require legislative initiatives to eliminate impediments to the exchange of information. Cultural barriers that have traditionally separated communities of managers from those of researchers must be broken down. This will require changes in modes of communication, as well as changes in reward systems for both communities. Successful ecosystem management depends on institutions that are adaptable to variations and changes in ecosystem characteristics, as well as changes in our knowledge base.

Finally, institutional barriers to learning can limit our capacity to reduce uncertainty (Lee 1993). For example, management agencies may lack systematic plans for learning that include prioritized listings of identified uncertainties, methods for reducing important and tractable uncertainties, procedures for evaluating existing actions, and mechanisms for retaining new knowledge in the memory of the institution (Hilborn 1987, 1992).

References

Bormann, F. H. and G. E. Likens. 1979. *Pattern and Process in a Forested Ecosystem.* Springer Verlag, New York, NY.

Botkin, D. B. and M. J. Sobel. 1975. Stability in time-varying ecosystems. *American Naturalist* 109:625–646.

Carpenter, R. A. 1980. Using ecological knowledge for development planning. *Environmental Management* 4:13–20.

Christensen, N. L. 1994. Plants in dynamic environments: Is "wilderness management" an oxymoron? In P. Schullery and J. Varley, *Plants and the Environment.* National Park Service. In press.

Christensen, N. L., A. M. Bartuska, J. H. Brown, S. Carpenter, C. D'Antonio, R. Francis, J. F. Franklin, J. A. MacMahon, R. F. Noss, D. J. Parsons, C. H. Peterson, M. G. Turner and R. G. Woodmansee. 1996. The scientific basis for ecosystem management. *Ecological Applications.* (in press).

Congressional Research Service. 1994. *Ecosystem Management: Federal Agency Activities.* Congressional Research Service, Library of Congress, Washington, DC.

Connell, J. H. and Sousa, W. P. 1983. On the evidence needed to judge ecological stability or persistence. *American Naturalist* 121:789–824.

DeAngelis, D. L. and Waterhouse, J. C. 1987. Equilibrium and nonequilibrium concepts in ecological models. *Ecological Monographs* 57:1–21.

Egerton, F. N. 1973. Changing concepts of the balance of nature. *Quarterly Review of Biology* 48:322–350.

Egler, F. E. 1964. Pesticides in our ecosystem. *American Scientist* 52:118–120.

Ehrlich, P. R. and A. H. Ehrlich. 1991. *Healing the Planet: Strategies for Resolving the Environmental Crisis.* Addison-Wesley Publishing Company, Reading, Massachusetts.

Franklin, J. F. 1993. Lessons from old-growth. *Journal of Forestry* 91:10–13.

Golley, F. B. 1993. *A History of the Ecosystem Concept in Ecology.* Yale University Press, New Haven, CT.

Government Accounting Office. 1994. *Ecosystem Management: Additional Actions Needed to Adequately Test a Promising Approach.* United States General Accounting Office Report to Congressional Requesters. GAO/RCED-94-111.

Hilborn, R. 1987. Living with uncertainty in resource management. *North American Journal of Fisheries Management* 7:1–5.

———. 1992. Can fisheries agencies learn from experience? *Fisheries* 17:6–14.

Holling, C. S. 1978. *Adaptive Environmental Assessment and Management.* John Wiley & Sons, New York, NY.

———. 1993. Investing in research for sustainability. *Ecological Applications* 3:552–555.

Keystone Center. 1991. *Biological Diversity on Federal Lands.* The Keystone Center, Keystone, CO.

Lee, K. N. 1993. *Compass and Gyroscope: Integrating Science and Politics for the Environment.* Island Press, Washington, DC.

Levin, S. A. 1992. The problem of pattern and scale in ecology. *Ecology* 73(6):1943–1967.

Likens, G. 1992. *An Ecosystem Approach: Its Use and Abuse.* Excellence in Ecology, Book 3. Ecology Institute, Oldendorf/Luhe Germany.

Lubchenco, J., A. M. Olson, L. B. Brubaker, S. R. Carpenter, M. M. Holland, S. P. Hubbell, S. A. Levin, J. A. MacMahon, P. A. Matson, J. M. Melillo, H. A. Mooney, C. H. Peterson, H. R. Pulliam, L. A. Real, P. J. Regal and P. G. Risser. 1991. The Sustainable Biosphere Initiative: An ecological research agenda. *Ecology* 72:371–412.

May, R. M. 1973. *Stability and Complexity in Model Ecosystems.* Princeton University Press, Princeton, NJ.

O'Neill, R. V., D. L. DeAngelis, J. B. Waide and T. F. H. Allen. 1986. *A Hierarchical Concept of Ecosystems.* Princeton University Press, Princeton, NJ.

Richardson, C. J. 1994. Ecological functions and human values in wetlands: a framework for assessing forestry impacts. *Wetlands* 14:1–9.

Slocombe, D. S. 1993. Implementing ecosystem-based management. *BioScience* 43:612–621.

Society of American Foresters. 1993. *Sustaining Long-term Forest Health and Productivity.* Society of American Foresters, Bethesda, MD.

Sutherland, J. P. 1974. Multiple stable points in natural communities. *American Naturalist* 108:859–873.

Tilman, D. and J. A. Downing. 1994. Biodiversity and stability in grasslands. *Nature* 367:363–365.

Turner, M. G. (ed.) 1987. *Landscape Heterogeneity and Disturbance.* Springer-Verlag, New York, NY.

Vitousek, P. M. 1990. Biological invasions and ecosystem processes: toward an integration of population biology and ecosystem studies. *Oikos* 57:7–13.

Walters, C. J. 1986. *Adaptive Management of Renewable Resources.* MacMillan, N.Y.

White, P. S. 1979. Pattern, process, and natural disturbance in vegetation. *Botanical Review* 4:S229–299.

Wilson, E. O. 1992. *The Diversity of Life.* Belknap Press of Harvard University Press, Cambridge, MA.

Notes

1.   Ecologists' use of the terms "goods" and "services" is not exactly congruent with usage among economists. For example, there are services such as legal services in the economy that are easily privatized and exchanged in markets and some goods such as roads that are not.

2.   The exact fate and number of each cell can be predicted with near absolute certainty in some organisms such as roundworms, but with much less certainty in organisms such as trees or humans.

3.   This is neither to deny nor diminish the importance of the evolutionary impact of organisms on one another that results in considerable integration of activity or coevolution among species.

CHAPTER TWO

Ecosystem Valuation: An Overview of Issues and Uncertainties

Michael A. Toman[1]

Introduction

The idea of "valuing ecosystems" has cropped up with increasing frequency in debates about the management and preservation of the resources and systems with which nature endows us. Concerns about ecosystems have become more prominent with the growing belief that threats to natural integrity and human well-being are increasing in scale, from those involving local systems (one lake) to effects on larger systems (a whole watershed) whose consequences may be both greater and longer-lasting. Concerns about the economic valuation of ecosystems in this context reflect a widespread interest within the policy process for more information with which to judge the severity of these threats, relative to other pressing private and public concerns.

To more than a few ecologists and other students of ecosystem management issues, this idea of valuing ecosystems seems a bit strange, if not bizarre. Isn't it self-evident that well-functioning ecosystems are valuable to us, that they are something we cannot do without? And how could one hope to put a price tag on an ecosystem, especially since it is in many respects the antithesis of a market good? Critics see the entire approach of economic valuation as too anthropocentric, with insufficient attention to the intrinsic worth of ecosystems, as well as dubious in practice because of the assumptions needed to carry out the exercise.

The attention to ecosystem valuation reflects at least in part the force of the economic model of comparative valuation and tradeoffs in policy analysis. From this perspective the views expressed in the previous paragraph are unsatisfactory because they do not provide a basis for setting priorities in ecosystem maintenance and protection. While few if any economists would argue that benefit-cost calculations can tell decision makers everything they need to know to allocate scarce societal resources, advocates see such calculations as providing highly valuable information. Expenditures that can generate more benefits or avoid more unwanted costs are candidates for increased resource allocations. According to this decision-making model, ecosystem valuation provides a means for assessing how ecosystem damages compare with other problems society must face, and a means for evaluating potential responses to damages. The point is not to dispute the status of ecosystems as having value outside the marketplace, to require economic proof that ecosystems are valuable to people, or to calculate their total worth like an asset that can be traded on a commodity exchange. Instead, the point is to calculate the *incremental* value or cost of changes in the conditions of ecosystems to guide public decisions.[2]

If one accepts this perspective (and critics of economic valuation often do not, as noted below), then the issue is not whether to value ecosystems but how. From this point of view, the important questions include: What attributes of ecosystem states and processes does one attempt to include in valuation? How can social values of these attributes be inferred? What tradeoffs are possible or meaningful in assessing ecosystem values? Are there value criteria (yardsticks) other than money that provide useful (or even better) information?

The next section of the paper discusses in more detail the basic reasoning behind viewing ecosystems as sources of value that are not only important to society but appropriate targets for economic valuation. That section also outlines the basic theory of economic valuation that underlies ecosystem valuation. The following section briefly summarizes the tools available to economists for this task. Following this overview of the economic approach to ecosystem valuation is a section that attempts to summarize some of the main criticisms that have been lodged against it.

The concluding section of the paper synthesizes previous arguments to express a judgment about the utility of ecosystem valuation with economic methods. In brief, I find sweeping objections to it to be less than compelling. However, there are ecological, economic, ethical, and empirical uncertainties that highlight the tentative status of ecosystem valuation estimates, especially for large-scale, complex systems. These uncertainties suggest that other ecological and institutional information is likely to be important in evaluating policies to reduce threats to ecosystem integrity.

Basic Concepts Behind Ecosystem Valuation

The economic approach to ecosystem valuation is built upon the same basic anthropocentric, utilitarian foundations as the modern theory of economic valuation and benefit-cost analysis generally (see Kopp and Smith 1993 for further discussion). Ecosystems provide what economists call valued flows of services and thus, in theory, warrant the same treatment as other valued social assets. "Service flows" here refer to *all* the sources of value that people perceive the natural systems provide. In addition to resource consumption (e.g., oil or fish extraction and waste disposal),[3] the services include nonconsumptive uses (low-impact recreational visits) and "passive use" values that may have little to do with direct use of any kind, such as the satisfaction of believing that reasonably intact, well-functioning ecosystems are an important part of our cultural and spiritual heritage and are worth preserving regardless of one's direct exposure to these systems. The "maintenance services" provided by well-functioning ecosystems—water and nutrient recycling, storage of biodiversity, climatic stabilization—also can be seen as a form of passive use in that they are background services that typically lie outside any individual's direct use choices, yet damages to these services are clearly relevant to human interests.[4]

Given that one accepts the conception of ecosystems as sources of socially valued services, valuation of these services and thus of the ecosystem states needed to produce these services is the logical next step in the economic paradigm. Here the heart of the matter is the economic concept of *preferences and tradeoffs.* Individuals are assumed to have a capacity, however subjective, heuristic, and inarticulated, to judge the consequences to them of changes in the availability of ecosystem services relative to other things they value. Based on these preferences and the knowledge they possess about the resources in question, they can, in principle, express a view about how much they are willing to give up in terms of other consumption opportunities in order to have more investment in the maintenance of ecosystem services, or how much compensation would offset the loss of ecosystem services.[5]

The previous paragraph describes the demand function for ecological services. The last step in the chain of economic reasoning is an understanding of the processes by which ecosystem conditions—their states and processes—give rise to services valued by people, and what is required in terms either of remediation or foregone use to offset loss of ecosystem services (or increase them). Given this "production function," it is possible in principle to assess not only the valuation of changes in ecosystem services, but also the real economic costs of providing them.

It is worth emphasizing that, while the economic paradigm is anthropocentric rather than ecocentric, and utilitarian rather than based on inherent rights, it encompasses a broad range of values that proponents of ecosystem preservation advance on "intrinsic worth" grounds. Since there are philosophical drawbacks to the ecocentric approach (see, e.g., Norton 1984), in principle the capacity to address a wide range of preservation motivations within an anthropocentric framework is important. Moreover, utilitarian motivations for valuing ecosystem services can reflect a variety of altruistic as well as more narrowly self-serving motives. The heritage values mentioned above can be part of a larger interest in passing on an appropriate bequest to subsequent generations, in protecting the knowledge or cultural positions of other groups than our own in this generation (e.g., the preservation of indigenous occupations or healing practices), or in promoting the material well-being of those portions of the current generation that live in closer contact with natural systems (e.g., agrarian societies).

The development of this valuation framework and attempts to implement it might be seen largely as academic exercises if there were well-functioning markets for ecosystem services. Obviously, this is not the case and cannot be the case for many ecosystem services, though specific components related to resource extraction or recreational amenities can and do get traded in markets with varying degrees of efficiency.[6] To determine how much of the "public goods" provided by ecosystems *should* be supplied and how best to supply them, economists apply the economic paradigm through benefit-cost analysis to find the resource allocation that makes the net gain to society as large as possible, given whatever physical and institutional constraints affect the supply of the public ecological goods.

There are well-known drawbacks to benefit-cost analysis as a normative policy analysis tool. In particular, the standard framework focuses on allocations that maximize a measure of aggregate value without concern for the distribution of benefits and costs, despite the self-evident social concern with distributional issues in policy debates.[7] Other concerns are related to the reliability of the valuations themselves. However, *some* kind of decision criterion is needed once one decides that the management and protection of ecological services is a public policy issue, i.e., within the realm of collective or state activity. Critics of the utilitarianism reflected in benefit-cost analysis suggest other criteria, as discussed below, and in determining what criteria seem to be most useful one must weigh the strengths and weaknesses of benefit-cost analysis against other approaches.

Several other points are important to bear in mind in considering this conceptual framework for ecosystem valuation. First, as noted above, the emphasis in the economic paradigm is on *incremental*, or *marginal* changes in ecological services. It is not on the total value of ecosystem services, let alone total ecosystems. Second, the emphasis in valuation is on benefits to humans as opposed to the systems themselves. For example, the emphasis is on the benefits derived from wheat consumption, not from the existence of nitrogen-fixing bacteria that make growing wheat possible.[8] Well-functioning ecosystems are important means to (numerous) ends.

A third point following from this emphasis on services is that the reliability of valuation estimates or the meaning that can be extracted from them, depends on what the service is seen to be by those whose values are being explored. For example, the results of a public opinion survey on environmental protection will depend on the kind of information people possess about the issue, the kind of information provided by the survey, and the context provided by the questioner. Broader or more diffuse ecosystem attributes will elicit different expressions of valuation than more narrowly defined ones: better understood attributes will be valued differently than those that are less well understood. This should not be surprising, since it mirrors our own experience with more conventional economic goods, but it is important to keep these points in mind when gauging the strengths and weaknesses of economic valuation tools.

Finally, the content of the economic approach to ecosystem valuation depends crucially on what one assumes about substitutability among ecosystem services or between ecosystem and other services. As noted above, the economic model is based on the assumptions that such tradeoffs are feasible and meaningful for the people facing them. Among the main criticisms of the economic approach are the points that substitution of other services for the functioning of ecosystems may be a dubious proposition, at least in some important circumstances, and that some changes in ecosystem impacts are not directly commensurable with other economic values because they involve overarching moral issues. I return to these points in the section entitled "Potential Criticisms of Economic Valuations."

Economic Valuation Methods

It is beyond the scope of this paper (and well beyond the writer's expertise) to provide a detailed review of techniques developed by economists for valuing nonmarket ecological services.[9] Thus, this section provides only a brief summary of the approaches and their capacities (see Braden and Kolstad 1991, Freeman 1993, and Kopp and Smith 1993 for thorough discussions of theory and applications).

The distinction between direct use and passive use values was introduced in the previous section. Within the first category of nonmarket values, perhaps the most common technique is known as the *travel cost method*. The name reflects the origin of the technique: Information on the willingness of individuals to travel different distances and incur different costs (including costs of time devoted to the activity) was used to construct a demand curve—a schedule of "willingness to pay"—for recreational experiences, even when these experiences had little or no direct price (e.g., park entry fee) attached. The technique can be used in any cases where consumption of an ecological service is associated with related activity in other markets from which information about willingness to pay can be constructed. In practice, most of the environmental applications have had something to do with recreation or aesthetics, including game preservation, visibility, and water quality. To apply the technique, it is necessary to survey users of ecological services in order to gather the necessary information about their expenditures and attributes.

So-called *hedonic methods* also have been used extensively in the environmental valuation literature. The basic idea behind these methods is that one can infer environmental values from the spread of market prices reflecting different "qualities." For example, differences in housing prices between clean and dirty airsheds (after controlling for other influences) can be interpreted as reflecting willingness to pay for better air quality; similarly, housing price differences before and after a toxic substance release can indicate willingness to pay to avoid the hazard. The application of hedonic methods to ecological services other than environmental quality is somewhat harder to envision since there are not well-organized markets for the related goods from which valuations can be directly inferred.[10]

The application of these methods is far more complex than these simple descriptions would indicate. Since the observed behavior in related markets will reflect economic agents' perceptions of the ecological attributes at issue, some understanding of these perceptions is needed to infer what is being valued. The greater is individuals' knowledge of the ecological services under consideration, and the less there is confounding ecological uncertainty, the clearer is the information the valuation studies provide. Uncertainty also clouds valuation since people's behavior may express both a reaction to an expected change in ecological services and a reaction to changes in risk per se. Finally, since many changes in ecological services may have a long duration—at the limit, they can be physically irreversible—hedonic or travel cost valuation estimates at a point in time must be augmented with assumptions about future valuations in order to paint a complete picture. In particular, a physically irreversible change must be valued according to the present value of the entire stream of future services foregone (Krutilla 1967; Krutilla and Fisher 1985).[11] This leads in turn to vexing questions about the appropriate discount rate to use for the present value calculation, a point discussed further in the next section.

Time and uncertainty provide a good point of departure for the discussion of passive use and nonuse values. The travel cost and hedonic methods provide ex post estimates of valuations based on actual behavior (revealed preference) in markets for goods related to the ecological services. Even in a static and deterministic setting, it is plausible to suppose that some individuals place a value on the existence of ecological services and attributes for their own sakes, quite apart from any consumption. Passive use values become even more complex when time and uncertainty are introduced, since there may be additional values attached to preserving the option of future use by oneself, one's peers, or future generations (Krutilla 1967; Krutilla and Fisher 1985). These values reflect the fact that preservation to maintain options is quite compatible with economic efficiency.

By their nature, ex post use-related valuation measures cannot provide a full accounting of these values. The only strategy so far developed by economists is the *contingent value (CV) method* (see Mitchell and Carson 1989 for a comprehensive review). In this method, surveys are conducted to provide information to respondents about some change in ecological conditions (e.g., loss of some identified species or some specified degradation of a site), and then to ask respondents to provide a statement of what they would be willing to pay to avoid a potential degradation or to ameliorate an actual degradation. The contingent values thus expressed will encompass both use values and nonuse values, given the description of the situation provided by the survey instrument; indeed, the estimates probably are more reliable if no effort is made to decompose somewhat artificially total value into components.

The CV method has a significant and growing track record of use. Nevertheless, it remains highly controversial among economists as well as between economists and noneconomists.[12] The controversy reflects a mixture of technical debate about practices that are still being refined; more fundamental concerns about the use of contingent survey information in lieu of revealed preferences based on actual market behavior; and the stakes that surround the application of CV in public and privately initiated litigation to compensate for natural resource damages.

Much of the concern about CV seems to reflect a basic distrust of contingent responses, based on the belief that individuals will not take the exercise seriously, will not understand the issues being addressed, or will introduce extraneous or strategic elements into their responses. Some of the concern also reflects a perception that the method has not performed well in practice because of apparent inconsistencies or paradoxes in results.[13] In contrast, proponents of the method argue, based on growing experience with the method and comparisons with results from other approaches, that while these are serious concerns, all of them are manageable with proper care; and that apparent inconsistencies in CV results reflect either design faults or misinterpretations of the evidence.

As proponents themselves note, the reliability of CV survey results depend crucially on how information is presented to respondents (the context), how well it is understood, how the survey is conducted, and how valuations are asked (e.g., a referendum on a specified hypothetical tax versus an open-ended valuation question). The greater the familiarity with the object or service being valued, and the greater the level of experience with valuing it, the more meaningful the results. The survey provides an opportunity for controlling the quality of information possessed by the respondents, in contrast to ex post valuations, in particular by explaining what is being valued and providing information that can help respondents in forming estimates of their valuations.

However, the challenge of valuing some ecosystem services, such as life support functions or biodiversity, or of valuing a complex package of services that ecosystems provide, is likely to be much greater than for valuing more familiar components of ecosystem services like recreation, aesthetics, survival of specific species, or pollution loadings. A very large amount of information that may be relatively remote from everyday experience must be conveyed to respondents in such cases. Moreover, the information must be conveyed in a way that reflects the substantial scientific or technological uncertainties surrounding ecosystem services. Finally, the information must also reflect a specified time frame so that people can distinguish, for example, an irreversible change from a long-term but reversible one.

It seems safe to say that even if one accepts the validity of the CV method for valuing specific goods or attributes, accurate estimates require careful (and possibly expensive!) survey design and administration. The importance of these methodological attributes only grows when one is valuing complex, unfamiliar ecosystem services.[14] On the other hand, to make no effort to value passive use or nonuse values is to give up on the possibility of putting an economic value on what may well be a key part of our stake in ecosystem services. In this case, it is necessary to consider what else might be done to assess our interests in ecosystems. The next section of the paper briefly considers some other possibilities that have been proposed, as well as examining various critiques of the economic approach.

Potential Criticisms of Economic Valuations

Before turning to specific criticisms of the economic approach to ecosystem valuation, it may be useful to summarize some general criticisms of the economic method (see Toman 1994 for a more detailed review). These can be divided into issues of *intergenerational concern* and *substitutability*. The issue of intergenerational concern has been alluded to earlier. With a positive discount rate in standard benefit-cost analysis, the present value of future costs inflicted from ecosystem degradation (and thus the benefits of preservation) tail off fairly quickly. With *any* positive discount rate, the present value of costs one century ahead are pennies on the dollar. This might be of little concern if overall economic welfare were growing apace. However, if the future ecological costs of our actions are thought to be substantial, and if prospects for remediation or compensation are limited, then there are significant ethical concerns associated with the standard approach (see Kneese and Schulze 1985 for further discussion).

The issue of intergenerational concern links with the substitutability issue at this point in the argument. In contrast to the picture painted above of ecosystems as "service factories," ecologists see these systems as complex, dynamic sets of processes that are organized on multiple scales (see Allen and Starr 1982, Holling 1986, Common and Perrings 1992, and Norton and Ulanowicz 1992). Smaller-scale components or subsystems (e.g., a single lake) respond quickly to stimuli and can recover relatively quickly from shocks; moreover, there is much more redundancy at this scale. For these reasons, the substitution paradigm in economics can fit well with the function of lower-scale ecological systems as sources of services. In contrast, higher-scale systems (e.g., entire watersheds; in the limit, the biosphere as a whole) respond much more slowly and with less redundancy. While even large-scale systems are resilient up to a point, it is possible to push them past that point and trigger rapid, discontinuous changes in function that may require large amounts of time for recovery. As a consequence of these features, the substitution paradigm that allows for incremental changes may be less well-suited to large-scale ecological impacts. This does not mean that the factory-of-services model is wrong, but it does suggest that the rules governing the factory can be quite different from those typically assumed in economic analysis.

With these thoughts in mind, we can group the most important objections found in the scholarly literature into three somewhat overlapping categories: (A) Values based on preferences are inherently irrelevant as a guide to valuation and policy; (B) Values based on preferences inherently are inadequate for these tasks; and (C) Other information is more useful as a guide to ecosystem evaluation for both scientific and policy purposes. Each of these points of view is briefly discussed in turn below.

Values Based on Preferences
Are Inherently Irrelevant

Perhaps the strongest advocate of this view is Sagoff (1994), who provides a caustic critique of economic axioms as well as practices, though his views are shared by a number of others who are critical of neoclassical economics as a guide to ecological decisionmaking. Sagoff's primary objections are that (1) preferences, as he conceives them, do not drive behavior; and (2) preferences do not imply well-being. He attempts to substantiate these points through a combination of logical argument, casual observation, and citation of experimental evidence that he interprets as indicating the inconsistency of economic valuation measures with theoretical predictions. He asserts that habits, history, ethical beliefs, and other such factors are more powerful tools for predicting behaviors. Sagoff also accuses economists of committing the logical error of inferring that assumed preference axioms are a correct representation of behavior, when in fact there could be many other explanations. In other words, he asserts, the preference model is impossible to prove wrong and therefore cannot be a scientific basis for explaining behavior. Finally, he asserts that important social virtues such as liberty are better served by basing policy decisions on the actual behaviors of people than on hypothetical estimates of what they prefer or value. Thus, he would argue, voting behavior in real political contests is sounder than contingent valuation analysis for making resource preservation decisions.

It may not be surprising that as an economist, I find much of this criticism less than compelling. It seems to me that preferences are part of the axioms of economics, and it is never possible to prove or disprove one's axioms. The key issue for economics is whether the predictions of the axioms stand up to empirical scrutiny. In particular, do people tend to consume less and produce more when commodities become more expensive? The performance of economics, while hardly unblemished, is pretty good when one considers the complex set of influences that must be untangled in the absence of laboratory experiments. The experimental evidence Sagoff cites as proof that economics does not explain behavior well is dated and seen by proponents of economic valuation as flawed. His casual empiricism seems no less ad hoc than the school of thought he proposes to attack.

This argument, if accepted, defends the usefulness of economic valuation based on observed behavior, but it leaves open questions about the reliability of contingent valuation. Here Sagoff does have a point in that even if the results of CV studies were accurate, society could insist on a different and more concrete standard of proof for calculating values or damages. However, proponents of CV note that its results tend to be corroborated by other methods (e.g., travel cost analysis) in cases where both methods have been applied to value the same good and the CV study has been designed and implemented well. This provides some indirect evidence anchored in real behavior for the usefulness of CV. The point is not that CV is perfect—it clearly is not—but that it provides useful information about values whose size cannot be computed through other economic methods. To eschew economic valuation entirely, as Sagoff proposes, and rely instead only on observed economic and political choices is to either ignore what appear to be important social values, or to assume that observed political and social decisions are more unimpeachable indicators (i.e., no "political market failures").

Values Based on Preferences Are Inadequate

This criticism of economic valuation applied to ecosystems does not try to deny the conceptual validity of economic valuation based on preferences, but rather to argue that individual preferences as expressed by willingness to pay cannot fully address the importance of large-scale ecosystem integrity to human society. Different versions of this argument have been made by a number of economists and philosophers; see, e.g., Norton (1982, 1992), Randall (1986), and Page (1983, 1991, 1992). The basic point is a concern that simple aggregations of individual willingness-to-pay assessments cannot adequately capture the moral dimensions (in particular, the impacts on future generations) of large-scale threats to natural integrity. In effect, this view takes issue with the standard assumption that all values that matter are commensurable (in technical terms, that preferences provide a *complete* ordering of all options). The solution proposed is not a rigid set of rules mandating ecosystem preservation as a good in itself that trumps all other values.[15] Instead, the argument is that willingness-to-pay, tradeoff-based measures of value must be supplemented with more "constitutional" societal rules or moral guidelines to reflect the moral issues inherent in preserving ecosystem integrity at large scales. This two-tier approach to governing of ecosystem conditions is often referred to as a *safe minimum standard* (see Toman 1994).

The determination of these standards can and should reflect the full range of interests of members of the society who are imposing the standards on themselves, including interests reflected in standard economic valuation estimates. Unlike point (A) above, there is an important role for preference-based analysis even with a safe minimum standard. It is the frame of reference that distinguishes the determination of social standards from individual valuations (see especially Page 1992 for an elaboration of this point). The importance of the safe minimum standard approach remains a source of disagreement among economists and other analysts, since it could also be argued that concerns about the future or other ethical principles could be assessed through sufficiently detailed conventional valuation analysis.

An important illustration of these points can be given in considering the discounting of future returns in benefit-cost analysis. As noted above, conventional practice makes any impacts more than one generation in the future essentially irrelevant from the standpoint of current decisionmaking. One response to this could be a mandated lowering of discount rates in evaluating ecosystem impacts. However, if such valuations became implemented in policy, the resulting distortions of capital flows would cause significant loss of economic efficiency as well as environmental harm.[16] A better alternative within the standard economic paradigm is to recognize that ecosystem damages involve the bundling together of two values: concern about ecosystem services, and concern about the aggregate bequest made available to future generations. Explicitly recognizing this social bequest element allows for increased concern about ecosystem integrity without distorting investment (in fact, since people are presumed here to care about the future, overall well-being would be enhanced). The safe minimum standard would go further in the case of large-scale risks by establishing a prior restraint on the degradation of ecosystem integrity.

Other Measures Are More Useful

Even if one accepts the usefulness of preference-based economic valuation in principle. and especially if one has doubts about it, one can believe that the economic valuation measures discussed do not adequately or reliably describe the importance of ecosystem conditions in practice. One might hold this view in particular given the complexity of ecosystems outlined previously in the paper. In response to this complexity, one might search for more direct ecological indicators of a system's condition or integrity.

A number of such approaches are found in the ecology and ecological economics literature. One common strategy is to use energy as an internal unit of account and trace the amounts of direct and indirect energy flows required to carry out the functions being asked of ecosystems. including those being demanded by humans. More complex input-output models of ecosystem activity also can be employed.[17] Measures of ecosystem organization and resilience can be incorporated in overall indexes of ecosystem health (Costanza 1992).

Clearly, these ecological analyses can provide useful scientific information about the conditions of ecosystems that are germane to policy debates about ecosystem management. However, as Page (1992) somewhat disarmingly notes. all analyses of this type lack a crucial element for shedding light on the values that should be attached to ecosystem functions or services: a connection between the posited measure of ecosystem state or activity on the one hand, and a measure of human interest or welfare on the other, as in the homely example of wheat cultivation and nitrogen-fixing bacteria mentioned previously. The explicit or implicit assumption in ecosystem analyses of the type just mentioned is that there is a one-to-one correspondence between the posited measure (e.g.. ecological energy efficiency) and human well-being.[18] While this is probably true to a significant extent at a large-scale or global level. there is often more room for tradeoffs in actual management problems.

A different argument is made by Vatn and Bromley (1994). These authors raise no quarrels with economic value but question whether inferred prices provide reliable information for the valuation of ecological resources that are unfamiliar. In many respects their argument parallels those made in the CV literature regarding the need for care in designing the survey instrument to adequately communicate the commodity to respondents. However, their conclusion is different: They argue that societies develop a variety of other institutions, including public resource management subject to the political process, for expressing preferences, delegating authority, and economizing on information requirements for decisionmaking. In this respect their argument echoes some of the points made in the safe minimum standards literature. Even if one does not have their doubts about the information revealed by willingness-to-pay studies, their emphasis on the institutional (process) aspects of valuation and decisionmaking seems well-placed.

Concluding Remarks

The arguments in this paper suggest first that economic valuation of ecosystem services is an important tool for public policy analysis and ecosystem management. Relying solely on ecological information or fixed philosophical principles for these purposes seems unlikely to produce policy decisions that best meet social interests, or even decisions that are compatible with the individual interests of the body politic. Sweeping conceptual criticisms of economic valuation seem misplaced, and growing experience with economic valuation of ecosystem services provides growing confidence in the ability to use economic valuation to get useful information.

Having said this, it must also be stated that economic valuation methods become more and more difficult to apply as the services in question become more complex, large-scale, interconnected, and subtle. These appear to be exactly the attributes that surround a number of broad ecological management issues: the scale of forest, wetlands, or other ecosystem areas to leave intact, the reduction of carbon dioxide or other accumulative substances in various media. For such problems, there are large technical and scientific uncertainties. Moral concerns can loom large, as can concerns with how decisions are made. In these circumstances, application of economic valuation tools needs to be seen as an evolving art capable of providing useful information but not single-handedly resolving dilemmas. Ecological information and signals from political and social debates also are important to bring into the decision process. To use these information sources, value judgments on all sides need to be made more transparent. This is turn requires an increase in the maturity of the political and social debates that often surround ecological preservation and management issues.

References

Allen, T. F. H. and T. B. Starr. 1982. *Hierarchy: Perspectives for Ecological Complexity*. University of Chicago Press. Chicago, IL.

Arrow, K., R. Solow, P. Portney, E. Leamer, R. Radner, and H. Schuman. 1993. *Report of the NOAA Panel on Contingent Valuation*. National Oceanic and Atmospheric Administration. Washington, DC.

Braden, J. B. and C. D. Kolstad (eds.). 1991. *Measuring the Demand for Environmental Quality*. North-Holland. Amsterdam, The Netherlands.

Brennan, T. J. 1989. A methodological assessment of multiple utility frameworks. *Economics and Philosophy* 5(2):189–208.

Burtraw, D. and R. J. Kopp. 1994. *Cost-Benefit Analysis and International Environmental Policy Decision Making: Problems of Income Disparity*. RFF Discussion Paper 94-15. Resources for the Future. Washington, DC.

Cleveland, C. J. 1991. Natural resource scarcity and economic growth revisited: Economic and biophysical perspectives. In R. Costanza (ed.) *Ecological Economics: The Science and Management of Sustainability*. Columbia University Press. New York, NY.

Common, M. and C. Perrings. 1992. Towards an ecological economics of sustainability. *Ecological Economics* 6(1):7–34.

Costanza, R. 1992. Toward an operational definition of ecosystem health. In R. Costanza, B.G. Norton, and B.D. Haskell (eds.) *Ecosystem Health: New Goals for Environmental Management*. Island Press. Washington, DC.

Diamond, Peter A. and Jerry A. Hausman. 1994. Contingent valuation: Is some number better than no number? *Journal of Economic Perspectives* 8(4):45–64.

Freeman, A. M., III. 1993. *Measurement of Environmental and Resource Values: Theory and Methods*. Resources for the Future, Washington, DC.

Hanemann, W. M. 1994. Valuing the environment through contingent valuation. *Journal of Economic Perspectives* 8(4):19–44.

Hannon, B. 1991. Accounting in ecological systems. In R. Costanza (ed.) *Ecological Economics: The Science and Management of Sustainability.* Columbia University Press, New York, NY.

Hausman, J. A. (ed.). 1993. *Contingent Valuation: A Critical Assessment.* North-Holland, New York, NY.

Holling, C. S. 1986. Resilience of ecosystems; Local surprise and global change. In W.C. Clark and R.E. Munn (eds.) *Sustainable Development of the Biosphere.* Cambridge University Press, Cambridge, UK.

Kneese, A. V. and W. D. Schulze. 1985. Ethics and environmental economics. In A. V. Kneese and J. L. Sweeney (eds.) *Handbook of Natural Resource and Energy Economics,* vol. I. North-Holland, Amsterdam, The Netherlands.

Kopp, R. J. and V. K. Smith (eds.). 1993. *Valuing Natural Assets: The Economics of Natural Resource Damage Assessment.* Resources for the Future, Washington, DC.

Krutilla, J. V. 1967. Conservation reconsidered. *American Economic Review* 54(4):777–786.

——— . and A. C. Fisher, 1985. *The Economics of Natural Environments: Studies in the Valuation of Commodity and Amenity Resources* (2nd edition). Resources for the Future, Washington, DC.

Mitchell, R. C. and R. T. Carson. 1989. *Using Surveys to Value Public Goods: The Contingent Valuation Method.* Resources for the Future, Washington, DC.

Norton, B. G. 1982. Environmental ethics and the rights of future generations. *Environmental Ethics* 4(4):319–330.

——— . 1984. Environmental ethics and weak anthropocentrism. *Environmental Ethics* 6(2):131–148.

——— . 1989. Intergenerational equity and environmental decisions: A model using Rawls' veil of ignorance. *Ecological Economics* 1(2): 137–159.

——— . 1992. Sustainability, human welfare, and ecosystem health. *Environmental Values* 1(2):97–111.

——— . and R. E. Ulanowicz. 1992. Scale and biodiversity policy: A hierarchical approach. *Ambio* 21(3):244–249.

O'Neill, R. V., D. L. DeAngelis, J. B. Waide, and T. F. H. Allen. 1986. *A Hierarchical Concept of Ecosystems.* Princeton University Press, Princeton, NJ.

Page, T. 1983. Intergenerational justice as opportunity. In D. MacLean and P.G. Brown (eds.). *Energy and the Future,* Rowman and Littlefield, Totowa, NJ.

——— . 1991. Sustainability and the problem of valuation. In Robert Costanza (ed.). *Ecological Economics: The Science and Management of Sustainability.* Columbia University Press. New York. NY.

——— . 1992. Environmental existentialism. In R. Costanza. B.G. Norton. and B.D. Haskell (eds.) *Ecosystem Health: New Goals for Environmental Management.* Island Press. Washington. DC.

Pollak. R. A. 1978. Endogenous tastes in demand and welfare analysis. *American Economic Review* 68(2):374–379.

Randall. A. 1986. Human preferences. economics. and the preservation of species. In B.G. Norton (ed.) *The Preservation of Species.* Princeton University Press. Princeton. NJ.

Sagoff. M. 1994. Should preferences count? *Land Economics* 70(2):127–144.

Simpson. R. D.. R. A. Sedjo. and J. W. Reid. 1994. *Valuing Biodiversity for Use in Pharmaceutical Research.* Paper presented to the AERE Workshop. "Integrating the Environment and the Economy: Sustainable Development and Economic/Ecological Modelling." June 6.

Toman. M. 1994. Economics and "sustainability": Balancing trade-offs and imperatives. *Land Economics* 70(4):399–413.

Ulanowicz. R. 1991. Contributory values of ecosystem resources. In R. Costanza (ed.) *Ecological Economics: The Science and Management of Sustainability.* Columbia University Press. New York. NY.

Vatn. A. and D. W. Bromley. 1994. Choices without prices without apologies. *Journal of Environmental Economics and Management* 26(2): 129–148.

Notes

1.   I am grateful to Anna Alberini and David Simpson for comments on earlier drafts. Responsibility for the opinions expressed in the paper is solely mine.

2.   Incremental or marginal analysis is a core element of the economic model; it is the way analysts determine whether greater net benefits can be obtained if some additional resources are allocated toward or away from different activities. The focus on incremental values does reflect important assumptions. as discussed below.

3.   When we dispose of materials or heat that we do not want into the natural environment. we are consuming the biosphere's capacities to render these effluents less deleterious to our interests. As with fish

and oil, these waste disposal services may be renewable (within limits) or nonrenewable (within a realistic time frame).

4.   In practice, the division of values into direct use and passive use categories is not clear-cut (e.g., biodiversity provides direct use services for pharmaceuticals). Since total value may be most important for gauging ecosystem preservation interests, however, this ambiguity is not fatal.

5.   What I have very briefly summarized here is the entire theory of economic preferences as applied to market or nonmarket goods. See Freeman (1993) for a comprehensive presentation.

6.   The boundary between market and nonmarket goods is fluid, as indicated by the growing volume of market transactions in genetic resources (Simpson, Sedjo, and Reid 1994).

7.   See Burtraw and Kopp (1994) for a review of attempts to address these issues; Toman (1994) discusses intergenerational distributional issues.

8.   For a number of people, not just ecologists, the functioning of the food chain may itself be a thing of beauty; that is certainly not ruled out in the economic paradigm.

9.   I do not discuss here the valuation of market services, such as extracted resources. In practice the valuation of these resources is important in detecting the size of market failures. However, the techniques (e.g., analysis of production functions and demand functions to uncover marginal revenues and costs) are more straightforward (though not trivial), and the valuation exercise is not as central to the management problem (if one suspects market failures, institutional reforms may be recommended as well as or in lieu of corrective taxes).

10.   Given information about willingness to pay associated with different levels of cost and different consumer characteristics (e.g., income, education), it is possible under certain conditions to reconstruct from travel cost information not just the marginal willingness to pay but also the entire "utility function" describing preferences for the nonmarket good in question. With this information, an analyst can figure out the economic value of nonmarginal changes in the supply of the nonmarket good. This capacity is important since many public policy debates involve such nonmarginal changes, e.g., opening of a large wilderness area to development. Marginal willingness to pay will understate total value for nonmarginal changes since it ignores the "consumer surplus" reaped from the inframarginal, more highly valued units of consumption. In general, it is more problematic to recover the whole preference ordering from hedonic estimates, limiting their applicability to marginal variations.

11.   Another complication is the likelihood that preferences themselves evolve over time, and that these changes are at least partly in response to changes in circumstances, including the state of natural systems. Krutilla (1967) drew attention to the evolving nature of environmental preferences; see also Pollak (1978) for an effort to model endogenous preferences. In practice, the process of endogenous preference evolution is likely to involve a great deal of uncertainty, making deterministic or simple statistical models of less value.

12.   For a sense of the debate among economists, see Arrow et al. (1993), Diamond and Hausman (1994), and Hanemann (1994).

13.   Concerns about CV include: (a) responses are biased upward because of the hypothetical nature of the exercise—respondents do not seriously consider their budget constraints; (b) responses do not adequately vary with the scope of the damage scenario (the size of the damages presented); (c) individuals value a broader environmental good in their responses than was presented to them ("embedding"); and (d) responses are skewed by how surveys are presented. An additional problem arises in extrapolating the results obtained from a survey to a larger population, given that only a subset of the survey respondents may have had prior knowledge of the ecosystem damage—should only this fraction of the population be assumed to suffer adverse effects?

14.   As Ray Kopp has pointed out to me, there is growing experience in valuing complex ecological resources with CV that seem to indicate people's capacities to understand and meaningfully value the services at issue. Much of this material is not yet published because it is the result of natural resource damage litigation. The Interior Department and the National Oceanic and Atmospheric Administration are in the midst (as of early 1995) of promulgating regulations governing the use of CV in such cases. As experience with large-scale, state-of-the-art CV grows, it will be possible for both proponents and opponents of CV to ground their arguments with the new evidence.

15.   As Brennan (1989) points out, efforts to represent hierarchical values in multiple-tier preference orderings tend to founder logically, in that the construction of the tiers (what goes where?) is ad hoc.

16.   Reducing the cost of capital serves to only to raise the visibility of future damages; it also increases the attractiveness of capital-intensive investments that often bring environmental damage in their wake.

17.   For summaries of these approaches, see Hannon (1991), Ulanowicz (1991), and Cleveland (1991).

18.   For example, Norton (1989) asserts that no generation should destabilize the ecosystem functions that underlie and provide the context for all human activity.

CHAPTER THREE

Ecological Risk Assessment and Ecosystem Valuation

Steven M. Bartell

Introduction

Purpose

The main purpose of this paper is to explore potential relationships between ecological risk assessment and ecosystem valuation. The relationships may already be more than potential. For example, ecological risk assessments invariably provide information used in decision making, certainly in the environmental regulatory arena (e.g., RCRA/CERCLA, or Superfund). Decisions based in part on economic values assigned to ecological resources can determine the nature of the ecological objectives of the risk assessment. Such valuation, or the economic consequences of possible ecological impacts, might also dictate the degree of accuracy and precision required of the risk assessment. Similarly, economic considerations of the cost of additional ecological data might weigh the value of such new data in the overall decision process. These reasons alone argue for examining the interconnections between ecological risk assessment and ecosystem valuation.

An ecological risk can be defined as the conditional probability of the occurrence of an undesired change in the biophysical environment combined with an evaluation of the consequences of that change. Ecological risk assessment (ERA) has become an important conceptual and operational methodology used to estimate the probable impacts of human and naturally induced disturbances of nature (USEPA 1992). ERA is a logical extension of basic ecological disturbance theory and more traditional environmental impact assessment used in regulatory compliance (e.g., the National Environmental Policy Act, NEPA). ERA remains an embryonic and evolving science with emphasis directed thus far toward developing scientifically defensible capabilities for estimating the probabilities of different ecological impacts in relation to physical, chemical, and biological stressors (e.g., toxic chemicals). Demonstrably less effort has been invested in evaluating the consequences of such undesired ecological events should they occur.

The valuation process attempts to determine both use and non-use monetary values of ecological resources. Economic values derived from the direct use of ecological resources have become readily quantifiable in traditional markets, particularly for such renewable resources as timber and fisheries. Convincing arguments can be made that all real wealth derives from the direct use of natural resources (Mollison 1990). Alternatively, estimating non-use values continues to challenge resource economists and ecologists alike. Methods controversial in comparison to direct use valuation have been put forward to measure non-use values, including such methods as the contingent valuation process. Methods based on travel and tourism spending to visit parks and other scenic natural places have been used to estimate non-use values. The hedonic pricing system has also been advocated. The calculus of these methods are based in traditional market economics.

Road Map

The examination of relationships between risk and valuation begins with a brief overview and history of ecological risk assessment. The paper continues with a discussion of current approaches to assessing ecological risks, identifies some critical risk assessment issues, then turns to different models of ecological valuation. The discussion addresses ecological risk assessment in the context of more traditional economic valuation as well as in relation to more recent assessments of ecological values from the viewpoint of sustainable environmental management (or applied ecosystem management). The importance of different models or world views of nature on assessment and valuation are examined. Risk assessment and valuation are subsequently integrated in the context of sustainable environmental management. The chapter concludes with speculations concerning future developments in risk assessment and valuation of ecological resources.

Ecological Risk Assessment

Ecological risk assessment (ERA), as currently practiced (e.g., Bartell et al. 1992, Suter 1992), uses methods of systems analysis to integrate aspects of biology, ecology, environmental chemistry, environmental toxicology, hydrology, and other earth sciences to estimate conditional probabilities of the occurrence of undesired ecological events. In theory, ERA applies to both natural and human impacts on ecological resources. In practice, nearly all of the assessments address ecological impacts resulting from human activities.

In concept, *risk* refers properly to the conditional probability of a specified event occurring (e.g., dam failure, reactor meltdown, bridge collapse, plane crash) combined with some evaluation (e.g., a loss or damage function) of the consequences of the event (e.g., injury, death, excess cancer, property loss). Kaplan and Garrick (1981) introduced a cogent conceptual model of risk as an ordered triplet (see also Helton 1993):

$$\text{Risk} = f\ \{p_1, p_2, p_3\}$$

The first element of the triplet describes the nature of the event of interest. The probability of the event occurring designates the second element. The third member of the ordered triplet is an evaluation of the consequences of the event. Elements p_1 and p_3 are commonly interdependent. One of the considerations in determining "what is at risk" (p_1) stems from an implicit assessing of the consequences of the event (p_3), measured according to some value system. The consideration of consequences might determine the statement of p_1, generally or explicitly. The statement of p_1 may in turn largely dictate the data, methods, and models relevant for estimating p_2. This simple conceptual model for risk might also prove useful for pointing out risks with perhaps similar consequences but different sources (e.g., forest damage from acid deposition and pest outbreaks). The risk model might likewise underscore the need to consider tradeoffs in risk management.

By analogy to the model, an *ecological risk* is the conditional probability of a specified ecological event occurring, coupled with some statement of its ecological consequences. In nearly all applications, more than one ecological response will be of interest. The model logically extends to a vector of ordered triplets: considerations of time or space can add additional dimensions to produce a matrix of triplets.

The undesired ecological events constitute an unbounded set of negative ecological impacts, ranging in theory across the entire spectrum of ecological measures. More limited in application, these endpoints of risk estimation have routinely included unacceptable decreases in the abundance of one or more ecological populations of interest. The ecological consequences of undesired events that occur also lie within the domain of ecological risk assessment. However, assessments performed thus far have rarely addressed the ecological consequences of the disturbance or the probability (and rate) of recovery following disturbance. As currently practiced, however, assessing ecological risk essentially entails describing, either quantitatively or qualitatively, the likely occurrence of an undesired ecological event. The current US EPA (1992) framework for ecological risk assessment identifies *problem definition, exposure assessment, effects assessment*, and *risk characterization* as the necessary components of an ecological risk assessment. This framework extends beyond the probabilistic nature of ecological risk to include qualitative risks.

The USEPA (1992) framework for ecological risk assessment will be used to generate guidelines for ecological risk assessment in the U.S. This framework continues to evolve in relation to internal and external review. Consequently, the framework remains a working document and neither carries the force of law in the U.S. nor enjoys the official endorsement of the USEPA. Nonetheless, the framework represents the contribution of considerable effort on the part of ecologists, environmental toxicologists, chemists, and policy analysts, within and external to EPA personnel. Other similar approaches for performing ecological risk assessment have been developed (e.g., National Research Council, Barnthouse 1993; Water Environment Research Federation, Parkhurst 1993). However, these approaches overlap significantly with the USEPA framework.

The USEPA (1992a) framework identifies and describes in detail the four necessary components of an ecological risk assessment: (1) problem formulation, (2) exposure assessment, (3) effects assessment, and (4) risk characterization. Each component is briefly described in the following section.

Problem formulation identifies the nature of the ecological disturbance of concern (e.g., toxic substance) and the ecological resources potentially at risk. In formulating the problem, critical data are identified along with the relevant spatial-temporal scales of the assessment. Procedures are identified for integrating the data to address the selected assessment endpoints. An assessment endpoint is the p_1 in the Kaplan and Garrick model that answers the question, "what is at risk?" The EPA framework differentiates assessment endpoints from what can actually be measured or assessed, namely, measurement endpoints. For example, an *assessment endpoint* might be the risk of an unacceptable reduction in the local population size of a particular rare and endangered plant species. Depending on the availability of relevant and appropriate data, the measurement endpoint might only be correlated with the assessment endpoint. If the species of interest is rare and endangered, it is less likely that the stress-response relations needed for risk estimation will have been developed for the species of concern, and the assessment may be based on data developed for a related, but different, species. The surrogate species becomes the measurement endpoint and the utility of the risk assessment hinges on the validity of extrapolating from this related species to the rare and endangered species of interest. Ideally, the measurement and assessment endpoints would be identical; in practice, they should converge during the course of the assessment. Importantly, the problem formulation component of the assessment produces a comprehensive conceptual model that guides the overall ecological risk assessment.

Assessing exposure of the ecological endpoints to the disturbance constitutes the next phase of the assessment. Of interest are the magnitude, frequency, timing, and duration of the source of risk. These exposure estimates might result from direct measurements, model results, or extrapolations from laboratory and field experiments. The result is a profile of exposure that emphasizes quantitative measures of exposure concentrations (and associated uncertainties) for each of the assessment endpoints.

The *effects assessment* produces exposure-response functions for each of the ecological endpoints and stressors of concern. These functions are used to estimate the degree of ecological impact anticipated for various levels of exposure. These functions might result from direct measurements, laboratory and field experiments, interpolations and extrapolations using data from other similar substances and organisms, or from more complex ecotoxicological models. Establishing the uncertainties associated with the exposure-response functions is an important component of the effects assessment.

The process of *risk characterization* attempts to integrate the results of the exposure and effects assessment and estimate the ecological risks identified in the original problem formulation. Risk characterization may rely on direct measurements in the field, the results of experiments in laboratory or field systems (e.g., micro-, mesocosms), and estimates using various mathematical or statistical models. Combinations of these methods are frequently used in real-world assessments. A key component of risk characterization lies in describing and quantifying the uncertainties that entered into risk estimation and the significance of these uncertainties in the resulting risk estimates.

Previous assessments have focused on estimating ecological risks for different endpoints and disturbances (Barnthouse et al. 1982; O'Neill et al. 1982; Bartell et al. 1988; Cardwell et al. 1993, others). Increasing emphasis on risk to ecological resources in relation to RCRA/CERCLA enforcement, along with additional mandates such as the Natural Resources Damage Assessments (NRDA), is forcing valuation of ecological resources by risk estimators and risk managers. Nonetheless, assigning economic and cultural values to ecological resources at risk has historically been difficult using traditional economic models (e.g., Bateman 1993; Bateman and Turner 1993; Daly 1994; Erlich 1994; Peet 1992).

To develop an understanding of historical applications of ecological risk assessments, Bartell (1993) analyzed 126 reported assessments. The assessments were reconstructed from abstracts of ecological risk assessments reported at the annual meetings of the Society for Environmental Toxicology and Chemistry from 1980–1993. The analysis of these case studies examined the nature of the ecological stress, the primary medium of exposure, the level(s) of ecological organization assessed, the nature of the exposure, and the method of estimating risk. The results of this analysis demonstrated that ecological risk assessments, in practice, have focused on the potential impacts on populations of benthic invertebrates, fish, and zooplankton in relation to stresses imposed by trace metals, organic contaminants, and pesticides. The majority of the assessments relied on measured values of exposure or exposures estimated using transport and fate models (e.g., Calabrese and Kostecki 1992). Exposures were compared to toxicity data derived from laboratory bioassays. Approximately three percent of these assessments estimated *ecological risk* in probabilistic terms and nearly all of these were reported in the last three years.

Critical Issues in Ecological Risk Assessment

One of the challenges in examining the possible relationships between ecological risk assessment and ecosystem valuation is the fact that ecological risk assessment is an emerging discipline, in both concept and method. In assessing ecological risks, current quantitative capabilities in ecology and toxicology are quickly pushed to their limits (Suter 1981, 1992). Such limitations become evident in the process of risk estimation. Unacceptable risks can also result from the expected severity of the disturbance in question, from high uncertainty, or their combination (O'Neill et al. 1982). A respectable goal in advancing ecological risk assessment is to minimize the number of assessments of high risks that result primarily from large uncertainties and sparse information.

Several important issues must be successfully addressed to continue the advances made in the last decade of research, development, and application of ecological risk assessment. At least two of these issues are germane to the discussion of ecosystem valuation: (1) the identification of an underlying environmental model and (2) the continuing evolution of ecosystem concepts. The following sections address these two issues.

In Search of a Reference Environment

One critical problem facing ecological risk assessors and risk managers is the absence of an underlying ecological frame of reference. How society perceives, understands, and measures the natural world will influence the kind of human-environmental ensemble that we choose to sustain and which can become a justifiable baseline for assessing risk. Both ecological risk assessment and ecological resource valuation are influenced by particular natural world views (Holling 1994). The following discussion outlines three alternative "metaphors of nature": an equilibrium model, a dynamic model, and an evolutionary model (Holling 1986). Each has its own characteristic ecological, management, and valuation implications. These alternative models of nature also imply differences in human perceptions of the past and the future. Such differences in perception may importantly influence risk management.

An Equilibrium Model

The equilibrium metaphor of nature represents the ecological systems as constant in time and homogeneous in space. Nature is simply an inexhaustible environmental backdrop or context against or within which the human events unfold. This perspective leads to the conclusion of economies of scale—more and bigger are always better. This equilibrium model also includes the assumption of a resilient nature that returns predictably to some underlying, but unspecified, ecological *status quo* following the removal of the source of stress or disturbance.

The equilibrium viewpoint connotes several implications for ecological risk assessment and ecosystem valuation. The equilibrium or "nature is constant" metaphor suggests that we look to the past to determine endpoints for ecological risk assessment.

If nature is constant, then unacceptable changes in the population sizes of valued species might be reasonably used as objectives in assessing the impact of human and natural disturbances. Here, what is unacceptable is determined by a combination of ecological and societal concerns (e.g., criteria for endpoint selection in the USEPA Framework for Ecological Risk Assessment, USEPA 1992). This viewpoint might also identify tools for assessing risk. For example, the numbers or proportions of individuals in different age or size classes of a population (e.g., fish, birds, mammals) can be used to construct life tables that characterize the population. Changes in the distribution or abundance in the population life table that result from disturbance can be used to project the implications of the disturbance on future population size using matrix population models (Caswell 1989).

The equilibrium view of nature also connotes the valuation of ecological resources. The assumption of unlimited natural resources tends to decrease both the direct and non-use values of resources. Assumptions of continued availability (or substitution) might have been appropriate in the development of early economic models when human populations were comparatively small and external energy subsidies were not as well developed as in modern industrial society. Such assumptions, however, have likely played a significant role in determining the current human environmental and economic crises. If the full costs of resources were built into early economic models, it is probable that ecological risk assessment might never have been needed (or invented).

A Dynamic Model

A second model of nature recognizes that ecological systems change in time and space independent of human intervention. In the absence of disturbance, these systems progress through a series of predictable ranges of physicochemical and biological configurations. Disturbances, natural or human, might shift the location and subsequent trajectory of the system within the broad band of possible system states. If the disturbance is sufficiently large, the system might be driven to a new set of possible (and predictable) configurations that no longer overlap the previous (e.g., alternative stable states—see Allen et al. 1977, Bartell et al. 1992).

One logical management scheme consistent with this world view centers on attempting to maintain ecological systems within their broad bands of previously characterized configurations. Deviations of a system beyond previously measured (i.e., natural) bounds might be interpreted as a bellwether of disturbance, signaling the execution of management activities aimed at restoring the system to within some desired dynamic range of natural order.

An important implication for risk assessment is that risk will have to be assessed against a backdrop of natural ecosystem variability. Instead of using some constant measure of an ecological resource as a frame of reference for selecting endpoints, endpoints developed under this dynamic model will necessarily include some characterization of resource variability. If the endpoints in the assessment include effects on a highly variable component of ecosystem structure or function, a greater degree of disturbance will likely be required to discriminate such impacts from natural variability. It follows from this observation that statistically verifying the occurrence probability for highly variable ecological endpoints, e.g., p_2 in the Kaplan-Garrick model, could require more intensive and costly sampling or monitoring (e.g., Niemi et al. 1993). The dynamic metaphor suggests that methods and models used to assess risk incorporate the ability to address or propagate variability (e.g., stochastic matrix models, Monte Carlo simulation). This capability is certainly consistent with the definition of ecological risk as a conditional probability; the results of these modeling approaches can produce probabilistic estimates of risk (e.g., Bartell et al. 1992).

The dynamic model suggests that risk managers ought to view variability as natural and attempt to maintain the systems within measured ranges of behavior and state, rather than at some specified value. By adopting this model, we might attempt to manage the trajectory of future system states to maintain the range of variation determined by historical measurements.

Dynamic ecosystems might also influence economic valuation activities by forcing the consideration of relevant spatial-temporal ecological scales in determining direct or non-use values. Odum (1994) discusses the relationship between fluctuating renewable resources and economies at some length in the context of pulse stabilization in coupled economic-ecological systems. At issue are the relative rates of resource depletion and resource renewal in relation to the scales of human economic institutions and demographics. Resources replaced at the scale of decades (e.g., forests, fisheries) or intergenerational time scales will have different direct use values, optional use values, and perhaps even different existence or bequest values than more resilient resources.

An Evolutionary Model

A third metaphor recognizes not only the dynamic state of nature, but also nature's capacity to evolve. To continue the broad band analogy, novel dynamic yet predictable domains of system state may emerge as evolutionary responses to random events, as well as human or natural disturbance. Invoking this metaphor, human activities unfold within a dynamic environmental context that offers the promise of heretofore unknown ecological worlds. Evolution and extinction are the point-counterpoint exchange in this world. DNA emerges as the raw material of paramount value; biodiversity measures the never finished products of the evolutionary process. One logical management strategy framed in an evolutionary perspective aims at not compromising the adaptability of organisms, whether humans or nonhumans.

One implication for ecological risk assessment is that measures of evolution or adaptability ought to be included as key assessment objectives. The difficulties of quantifying "rates of evolution" for most metazoans suggests that measures of genetic diversity or other correlates (e.g., rate of growth Caswell 1989) of evolutionary capacity be examined for use in such assessments. Emphasizing an evolutionary model for ecological risk assessment suggests that the historical ranges of variation important under the dynamic model of nature might prove less useful in developing assessment endpoints.

An implication of this metaphor for resource valuation is that the capability to evolve remains the underlying source of continued real wealth, or direct value. Certainly, the increasing ability of scientists to augment and assist (even direct and accelerate) evolutionary change through genetic engineering requires consideration in ways perhaps not yet imagined for assigning economic value to evolutionary change.

In summary, the particular model humans use to perceive and describe their ecological and environmental context influences the development of ecological risk assessment in theory and in practice. The selected reference model also determines the economic values assigned to ecological resources.

Ecosystem—An Evolving Paradigm

A second key issue concerns how we translate our perceptions and understanding of nature into concepts and methods that produce useful risk assessments and define feasible pathways to sustainability. Ecological risk analysis and resource valuation can profit by adopting an ecosystem approach. Ecosystems are the fundamental units of basic ecology (Odum 1969; Smith 1975), ecological risk analysis (Bartell 1990; Bartell et al. 1992; O'Neill et al. 1982, 1983), and ecological sustainability. Importantly, ecosystem understanding continues to evolve from largely mechanical and deterministic concepts (Forbes 1887; Tansley 1935; Lindeman 1941; Hutchinson 1948; Odum 1969) to scale-dependent and hierarchical descriptions (Allen and Starr 1982; O'Neill et al. 1986; DeAngelis 1992). Nonetheless, delineating the boundaries of ecosystems remains a challenge in theoretical and applied ecosystem science. Boundary definitions determined by ecosystem function should be considered in addition to structural aspects of ecosystems. The early ideas of

Margalef (1968) for using minimal energy transfer across time and space to delineate functional ecosystem boundaries appear as useful today as when first offered.

Unfortunately, *ecosystem* has largely become synonymous with *habitat* in current prescriptions for assessing ecological risk (e.g., USEPA 1992). Significantly, the ecosystem as merely habitat has been historically downplayed in comparisons to the repeated emphasis of interactions and feedbacks among the physical, chemical, and biological components as the essential ingredient that distinguishes ecosystems (e.g., circular causal mechanisms, Hutchinson 1948) from habitat. One consequence of this emphasis on biotic-abiotic feedback is that ecosystem-level responses to disturbance should, by definition, be assessed as alterations of such interactions and feedback mechanisms. Assessment endpoints should be developed from considerations of ecosystem function.

Another critical issue in assessing ecological risks concerns the nature of mathematical or computer simulation models used to estimate risks. Nature challenges risk analysts with middle number systems that are inherently complex (Allen and Starr 1982; Weinberg 1975). The strength of different ecological perspectives (i.e., population, community, ecosystem) in model development lies in the nature of simplifying assumptions that delineate certain kinds of measurements and facilitate interpretation of new observations. Using different ecological concepts to their full advantage in ERA, particularly in identifying what is important to measure (i.e., endpoints) or model, requires greater ecological sophistication than that offered by the simple nested model. Current debates of simple versus complex models for risk assessment should be replaced by considerations of model adequacy. Simplifying our assumptions or simplifying our models does not simplify nature.

Ecosystem Valuation

Perhaps the most important aspect of ecosystem science in economic valuation is the increasing recognition of the monetary value of the functional services provided by ecological systems particularly in the processing of industrial wastes (i.e., assimilation capacity) and in the quality renewal of air, water, and soils that suffer the impacts of an increasingly industrialized society.

Assigning values to ecological systems can proceed from considerations of ecological values of significance and from economic values. Each perspective carries important, and different, connotations for risk and resource management. Ecological risk assessment may play differently into each perspective on ecological valuation. The next sections discuss different perspectives and approaches to assigning values to ecosystems and ecological resources in general.

Economic Perspectives

Natural resources are the ultimate source of real wealth (Mollison 1992). Assigning values to ecological systems from a basis in traditional economics leads to considerations of nature that differ from purely ecological values. With due apologies to deep ecologists, it might be argued that, from a strict ecological perspective, natural resources are value neutral. That is, without knowledge of future states of nature and considering the functional redundancy characteristic of ecological systems (i.e., Hill and Wiegert 1980), as well as the history of extinctions recorded in the fossil record, assigning ecological value to specific ecosystem components reflects more the ecology of the observer than the ecosystem. Nature simply is.

From an economic perspective, ecosystems are valued as natural resources rather than fundamental units of nature. The direct use values of ecological resources can be assigned in direct relation to a market economy. For example, King (1994) tallied the ex-vessel value of the 1992 Chesapeake Bay oyster harvest as $2.5 million, which generated business sales of about $7.5 million, household income of approximately $13 million, and combined taxes of another $4.6 million.

The primary approaches used in traditional economics to assign non-use values to ecological resources include the contingent valuation methods, the travel cost model, and the hedonic model. A detailed examination of each of these approaches lies clearly beyond the scope of this paper. Useful introductions and discussions from an environmental perspective include Jansson et al. (1994) and Kopp and Smith (1993).

One consequence of viewing ecological systems as *natural resources* is a conservation, rather than a preservation, ethic. One important implication of a conservation ethic is the eventual use or development of the resource. Resource conservation might result in problem formulations and assessment endpoints that differ from an ecological risk assessment developed from a preservation ethic. For example, acceptable reductions in a resource could define risk endpoints in relation to natural rates of resource renewal from a conservation perspective; the capacity for the resource will not be knowingly placed at risk, yet specific elements of the resource may be temporally sacrificed. In contrast, a preservationist approach to the same assessment might insist on zero risk to the current, as well as future, resource as the only defensible assessment endpoint.

Ecological Perspectives

Ecological entities (e.g., organisms, population, communities, ecosystems) are inherently value neutral. However, because natural systems have been observed to change in somewhat of an orderly fashion (i.e., successional trends), any particular ecosystem component might possess "ecological value" as a participant or effector of successional change or a future state of nature.

Assigning values from a purely ecological perspective focuses attention on ecological systems as fundamental units of nature. Establishing the value of ecological systems from an ecological perspective tends to focus attention on the natural structure and function of these systems. Several aspects or properties of ecosystems are inherently valuable and requisite for sustaining life (Jansson and Jansson 1994). The transformation of incident solar radiation to biochemical potential energy remains the foundation of natural capital. The resulting primary production base, combined with an associated consumer network, in addition to system metabolism, material recycling, waste processing, habitat generation, and the resulting diversity, resistance, resilience, and self-organization provide the basis for valuation and sustainability.

An ecological economics analysis of the Chesapeake oyster fishery might focus on the service provided by these animals in filtering the water of excess particulates. Before commercial fishing, the oysters effectively filtered the Bay every three to four days and maintained high water quality that translated into diverse biological assemblages. Current populations require nearly 300–400 days to filter the Bay. Thus, from a functional perspective, there has been a significant loss in the ecological value of the oyster populations in the Chesapeake Bay (King 1994).

A purely ecological approach to valuation leads to a preservation ethic for ecological systems. Without knowledge of the future states of nature, it is extremely difficult to assign traditional economic value to ecological resources. In a sense, ecological entities are not resources, but are inherently valuable and legitimate in their own right. This summarizes the valuation perspective from "deep ecology." From this viewpoint, a prudent policy in the face of economic (and ecological) ignorance is to preserve nature in all its pristine form whenever possible.

Ecosystem Valuation and the Law

Assigning values, either ecological or economic, to ecosystems or other natural resources does not occur in a legal vacuum. Laws are as much a component of society as economics and ecology. For example, legal precedents have been established for determining possible monetary compensation for environmental losses or damages in relation to the Natural Resource Damage Act (e.g., Ohio vs. the Department of the Interior). The NRDA essentially provides the trustee(s) of the resource compensation for injury or damages not redressed through the restoration and remediation processes of the Comprehensive Environmental Response, Compensation, and Liability Act of 1980, (CERCLA, or Superfund), and the Superfund Amendments and Reauthorization Act (SARA) in 1986.

Under CERCLA and the NRDA, terms like trustee, injury, compensation, and damages all have legalistic interpretation, developed though cases tested in the courts. A detailed presentation and analysis of the ecological and economic implications of the legal interpretations of these laws is beyond the scope of this report (and the competence of this author, who is not a lawyer). For a comprehensive treatment of the NRDA, see the volume edited by Kopp and Smith (1993). A few points are nonetheless worth mentioning. The NRDA has emphasized replacement cost as the monetary measure of economic damages to natural resources (e.g., Ohio vs. the DOI). Where restoration or other economic evaluations are included, the NRDA has commonly adopted the least cost alternative in awarding the compensation for damages to the trustees. Replacement has referred to the ability of the resource to provide environmental services equivalent in quality and quantity to pre-damage baseline status.

Ecosystem Valuation and Risk

There are at least four topics of discourse among ecological risk managers, risk assessors and ecological resource economists. These topics pertain to (1) developing meaningful objectives in assessing ecological risks, (2) defining necessary accuracy and precision in estimating risks, (3) determining the value of acquiring new information in the risk assessment process, and (4) characterizing the consequences of undesired ecological events. Each is presented in the following sections.

One, valuation of ecological resources, contributes to defining what is at risk in any assessment. Defining what is at risk is tantamount to delineating p_i in the Kaplan and Garrick ordered triplet. Thus, ecological and economic valuation are instrumental in selecting the objectives of the overall assessment. These interrelated values will be reflected in the development of the conceptual model that serves as the guideline for the risk assessment. Quite simply, ecological entities with high ecological or economic (or both) value should be selected as the objectives of the risk assessment. From an ecological perspective, dominant primary producers, important primary consumers, keystone predators, and critical decomposers, to the extent that they might appear sensitive to environmental disturbance, constitute meaningful objectives for ecological risk assessment. Economically, potentially sensitive species, ecosystems, resources, and habitats with significant direct and non-use use monetary values would similarly be justified as objects for risk assessment.

Two, the values of ecological resources at risk determine in part the degree of confidence necessary in the associated estimates of risk before the risk manager might make a decision. If the consequences are comparatively negligible, by whatever metric of value, the risk manager might justifiably select among management alternatives even in the face of considerable uncertainty. For example, if a local population of organisms appears at risk, but the population is characterized by a high intrinsic rate of increase or is surrounded by other populations that would provide a "seed" source for rapid recovery, the risk manager might feel secure using a qualitative, imprecise risk estimate to make a decision regarding the protection of the population. Alternatively, from an ecological valuation perspective, if the population is a rare or endangered species with a relatively low replacement rate and few nearby seed populations (by definition), the risk manager might reasonably insist on a rigorously quantitative and precise estimation of risk. Similarly, if the population is commercially valuable with either high use or non-use values, the risk manager might also insist on as rigorous an estimate of risk as possible before making a decision.

In practice, most ecological risk assessments address several objectives with varying ecological or economic value. Such values might assist in setting priorities for apportioning finite assessment resources to provide for maximum decision power per unit investment for each risk assessment objective (i.e., endpoint).

Three, ecological risk assessment has been developed as an iterative process (Bender and Jones 1993). The uncertainties that enter into risk estimation can result in the inability to discriminate among risk management alternatives, including taking no action. Under these circumstances, the economic or ecological (or legal) consequences of incorrect risk estimates might force another iteration aimed at reducing uncertainty to the point where a decision can be made. Ecological or economic values assigned to the objectives of the assessment can be used in combination with sensitivity/uncertainty analysis to identify the nature of new data that would provide the greatest return on investment in improving the risk estimates.

Four, all aspects of ecological and economic valuation of ecological resources might enter into characterizing the consequences of an undesired ecological event, should it occur. In addition, it is quite likely that such an evaluation will take place prior to selecting the event as an objective of the risk assessment, as suggested by the Kaplan-Garrick model (i.e., p_3). It might be reasonable to expect, following the occurrence of a risk event, that an even more comprehensive attempt at measuring the value of its impact, both ecologically and economically, would be carried out. Actual realization of the risk might reveal unforeseen aspects of resource value overlooked in the original development and selection of the event as an objective of the assessment.

Risk, Valuation, and Sustainable Environmental Management

Another motivation for this chapter is to offer sustainable environmental management as a logical and worthwhile context for assessing ecological risks, for developing and implementing plans for ecosystem management, and for calculating real values of ecological resources. Integrating ecological risk assessment with sustainable management will be discussed from alternative but complementary perspectives, that of larger scale government policy and its impact, and that of local or regional actions directed towards sustainable management. Underlying this motivation is the recognition that ecological risk assessment might be developed to the point of perfection, yet ecological resources will continue to degrade in the absence of a realistic economic context within which to value, assess, and manage ecological systems. The economics of sustainable environmental management might provide such context.

Simply stated, sustainable environmental management maintains the quality and quantity of ecological resources at approximate steady-state. Resources are depleted at rates not to exceed rates of their renewal. Depletion rate is measured over scales in time and space relevant to the resource and the user. Clearly, ecological resources cannot be sustainably managed without a prescription regarding quantity and quality. Thus, by embracing *sustainable environmental management* as an overarching management objective, decision makers and risk managers are forced to delineate, in operational detail, their underlying model for assigning ecological and economic values to natural resources. Risk assessors and managers could benefit from the precise statement of such a model, which would prove useful for developing meaningful assessment endpoints, identifying relevant information or data, selecting (or inventing) appropriate tools for risk estimation, and interpreting the ecological and economic consequences of risk.

Two approaches contribute towards developing and implementing sustainable environmental management practices. One, strategic national and international discussions, commitments, and treaties or protocols, can provide a foundation for future policy or regulations. Such larger-scale institutional actions may further encourage or compel subsequent state and local actions that are consistent with sustainability. Two, complementary to the incentives and policies promulgated by larger scale institutions are the actions of local and regional groups who seek to develop and implement sustainable environmental management plans without the insistence (or interference) of "big government." The tactic of these smaller scale movements is to force larger scale changes in response to local action. Both approaches need to be effectively combined to integrate risk assessment with sustainable environmental management. The strengths and limitations of each approach are briefly outlined in the following discussion.

Larger-scale approaches to risk management are certainly important and effective for managing resources that have characteristic spatial scales which exceed the jurisdiction of local or regional institutions. When local management activities impact upon resource availability at larger scales, institutions with jurisdiction at these larger scales are necessary to ensure equitable use and sustainability of the resource. For example, marine fisheries resources must be managed to guarantee sustained access by fisherman from many states. Equitable distribution of water rights in the arid western and southwestern United States remains a volatile political and economic issue requiring federal intervention. Other examples of top-down policies and activities that are consistent with the aims of sustainable environmental management include signing of the Montreal Protocols to decrease greenhouse gas emissions, participation in the 1991 Earth Summit meeting at Rio de Janeiro, and the recent formation of the President's Council on Sustainable Development.

One advantage of the large-scale approach is that resulting policy or legislation, once implemented, can have widespread impact in a relatively short time. One drawback is that large-scale policy may be inappropriate at the scale of impact. For example, regional differences in geochemistry are not included in the development of national water quality criteria; differences in water chemistry might result in standards too restrictive in some watersheds or not protective in others. Another drawback is the often lengthy process required to enact laws or develop policy at these larger institutional scales.

Regional differences in the distribution, abundance, and value of ecological resources underscore the validity of more localized approaches to sustainable environmental management that may prove more effective and more efficient than broad, sweeping, top-down policies for managing environmental risks. Regional planning has historically been an effective bottom-up approach to sustainable environmental management. The components of regional planning lend themselves for integration with the objectives and components of ecological risk assessment and ecosystem valuation.

Regardless of scale, a variety of environmental disciplines can contribute to integrating ecological risk assessment and ecosystem valuation with sustainable environmental management. Concepts and methods borrowed from landscape ecology, wildlife management, ecological restoration, and ecological engineering should be incorporated into consideration of waste site remediation. These disciplines can assist in defining meaningful assessment endpoints and in developing remediation goals that put remediation in a spatial context extending beyond the waste site boundaries. Ecological economics may inject the innovation in relations among ecological, legal, political, and social institutions required to implement plans for sustainable environmental management.

Summary Observations, Speculations, and Challenges

Ecological and Economic Scales

One difficulty in effectively addressing ecological concerns or intelligently managing ecological resources lies in the scale incompatibilities of ecological phenomena and human institutions entrusted with their stewardship. Ecological resources exhibit a diversity of characteristic temporal scales. Diurnal and lunar cycles are important to the growth and survival of some species. The life histories or production dynamics of many species are strongly seasonal. Yet for others, individual lives span decades, even centuries.

Businesses, banks, corporations, and other economic institutions are closely tied to quarterly reports and annual financial statements. Government institutions are active in relation to their annual fiscal cycles, as well as the two- and four-year cycles associated with elections. As a result of these periodicities and the nature of the political process, legislative institutions can introduce significant time lags into the risk management process. Relatedly, Holling (1986) observes that political institutions exhibit another temporal scale with a periodicity of 20–30 years. This period corresponds roughly to the turnover time of a generation of policy makers or regulators. This turnover permits re-evaluation of past policies and decisions by the following generation of politicos. Decisions can be overturned and policy revised with minimal ill will among personnel.

Appreciable differences in the activity patterns of regulatory or policy-making institutions and the ecological resources entrusted to or influenced by these institutions can pose nearly insurmountable problems in designing and implementing effective policy. One answer to the scale discrepancies is to permit the relevant time scale of the resource of concern to dictate the scale of the management activity. A recent Carnegie Commission report (1993) called for increased cooperation among federal agencies entrusted and empowered to maintain environmental quality. Risk assessment can provide a common operational framework to facilitate communication and cooperation. Transcending agency boundaries and jurisdictions can markedly and efficiently impact environmental quality.

Relocating the power to make decisions to local and regional institutions appears as another step towards reducing scale incompatibilities. As components of an integrated local or regional plan, ecological resources might be more realistically valued and effectively managed by those groups who know and understand the spatial-temporal dynamics of the resource in question and use them in a sustainable fashion at those scales. Instead of broad, sweeping management aimed at average circumstances (which might not exist), management and exploitation can be customized for the particular nature and distribution of the regional resources.

Challenges

There remain many challenges at the interface of risk assessment and valuation of ecological resources. We have made a value judgment that environmental sustainability is a necessary and desired management objective for society. Yet, the history of human civilization provides little evidence in support of sustainability. Virtually all of the cultures or societies that appear now or in the past to live within the resource constraints of their natural environs exist in near Stone-Age conditions or have been vanquished. Clearly, sustainability must be practiced at the global scale to ensure success.

The seminal work of Odum and Pinkerton (1957) points out thermodynamic limits to achievable efficiency in energetically open systems. Increased efficiency in energy and material use is a key component in sustainability. Therefore, inescapable thermodynamic limits on efficiency might preclude sustainability in the long run, regardless of the values and judgments of an enlightened society. History, thermodynamics, and the present world condition suggest we might be naive in entertaining sustainability. Nothing is forever, not even forever.

Sustainability need not be not the same as persistence. The maximum power principle suggests that society should strive for maximum energy throughput (Odum 1957). For it is the energy surplus above maintenance costs and the material by-products resulting from such surplus that fuel the development of society (e.g., arts, science, technology). Maximizing energy throughput provides for persistence as long as supplies of concentrated energy remain accessible at an energy cost less than the energy return in their utilization. History suggests that the maximum power approach is consistent, not with sustainability and living within natural resource constraints, but with expansionist policies and imperialism. Survival of the fittest?

In conclusion, this exploration into possible interrelations between ecological risk assessment and ecosystem valuation suggests that these endeavors are interdependent. The value of ecosystems (and ecological resources in general) can influence the selection of objectives in ecological risk assessment. The estimation of ecological risks includes evaluating the consequences, ecological and economic, of undesired ecological impacts, that is, ecological risk. Furthermore, the concepts of sustainable environmental management provide a meaningful context for risk assessment and ecosystem valuation. Yet, the process of assessment and evaluation cannot proceed unknowingly of legal precedents for risk assessment and valuation of natural resources. The relevance of ecological risk assessment and ecosystem valuation to each other may become clearer and more insightful with the continued conceptual and methodological evolution and advancement of both disciplines.

References

Allen, T. F. H. and T. B. Starr. 1982. *Hierarchy: Perspectives for Ecological Complex.* University of Chicago Press, Chicago, IL.

Allen, T. F. H., S. M. Bartell, and J. F. Koonce. 1977. Multiple stable configurations in ordination of phytoplankton community change rates. *Ecology* 58:1076–1084.

Baird, B .F. 1989. *Managerial Decisions Under Uncertainty.* Wiley-Interscience, New York, NY.

Barnthouse, L. W. 1993. Ecological risk assessment and the National Research Council. In Bender, E. S. and F. A. Jones (eds.) *Applications of*

Ecological Risk Assessment to Hazardous Waste Site Remediation. Water Environment Federation, Alexandria, VA.

Barnthouse, L. W., G. W. Suter II, and A. E. Rosen. 1990. Risks of toxic contaminants to exploited fish populations: Influence of life history, data uncertainty, and exploitation intensity. *Environmental Toxicology and Chemistry* 9:297–311.

———. D. L. DeAngelis, R. H. Gardner, R. V. O'Neill, G. W. Suter II, and D. S. Vaughan. 1982. *Methodology for Environmental Risk Analysis.* ORNL/TM-8167, Oak Ridge National Laboratory, Oak Ridge, TN.

Bartell, S. M., R. H. Gardner, and R. V. O'Neill. 1988. An integrated fates and effects model for estimation of risk in aquatic systems. *Aquatic Toxicology and Hazard Assessment: 10th Volume,* ASTM STP 971, American Society for Testing and Materials, Philadelphia, pp. 261–274.

Bartell, S. M. 1993. *Thirteen (+1) Years of Ecological Risk Assessment in SETAC.* Abstract of the Fourteenth Annual Meeting of the Society for Environmental Toxicology and Chemistry, November 1993, Houston, Texas.

Bartell, S. M., R. H. Gardner, and R. V. O'Neill. 1992. *Ecological Risk Estimation.* Lewis Publishers, Chelsea, MI.

Bateman, I. J. 1993. Valuation of the environment, methods and techniques: Revealed preference methods, pp. 192–265. In Turner, R. K. (ed.), *Sustainable Environmental Economics and Management—Principles and Practice.* Belhaven Press, New York, NY.

——— . and R. K. Turner. 1993. Valuation of the environment, methods and techniques: The contingent valuation method, pp. 120–191. In Turner, R. K. (ed.) *Sustainable Environmental Economics and Management—Principles and Practice.* Belhaven Press, New York, NY.

Bender, E. S. and F. A. Jones (eds.). 1993. *Applications of Ecological Risk Assessment to Hazardous Waste Site Remediation.* Water Environment Federation, Alexandria, VA.

Calabrese, E. J. and L. A. Baldwin. 1993. *Performing Ecological Risk Assessments.* Lewis Publishers, Chelsea, MI.

——— . and P. T. Kostecki. 1992. *Risk Assessment and Environmental Fate Methodologies.* Lewis Publishers, Boca Raton, FL.

Carnegie Commission. 1993. *Risk and the Environment.* US.

Cardwell, R. D., B. Parkhurst, W. Warrin-Hicks, and J. Volosin. 1993. Aquatic ecological risk assessment and clean-up goals for metals arising from mining operations, pp. 61–72. In, Bender, E. S., and F. A. Jones (eds.). *Applications of Ecological Risk Assessment to Hazardous Waste Site Remediation.* Water Environment Federation. Alexandria. VA.

Caswell. H. 1989. *Matrix Population Models.* Sinauer Associates. Sunderland, MA.

Daly, H. E. 1994. Operationalizing sustainable development by investing in natural capital. pp. 22–37. In. Jannson. A. M.. M. Hammer. C. Folke, and R. Costanza (eds.). *Investing in Natural Capital.* Island Press, Washington, DC.

DeAngelis. D. L. and L. J. Gross. (eds). 1992. *Individual-Based Models and Approaches in Ecology. Populations. Communities and Ecosystems.* Chapman and Hall, New York. NY.

Dobzhansky. T. 1968. Adaptedness and fitness. pp. 109–121. In R. C. Lewontin (ed.). *Population Biology and Evolution.* Syracuse University Press, Syracuse. NY.

Erlich. P. R. 1994. Ecological economics and the carrying capacity of the earth. pp. 38–56. In. Jannson. A. M.. M. Hammer. C. Folke. and R. Costanza (eds.). *Investing in Natural Capital.* Island Press, Washington. DC.

Forbes. S. A. 1887. The lake as a microcosm. *Bulletin Science Association of Peoria. Illinois.* Reprinted in *Illinois Natural History Survey Bulletin* 15:537–550. 1925.

Helton. J. C. 1993. Risk. uncertainty in risk. and the EPA release limits for radioactive waste disposal. *Nuclear Technology* 101:18–39.

Hill, J. and R. G. Wiegert. 1980. Microcosms in ecological modeling. pp. 138–163. In Giesy, J. P.. Jr. (ed.). *Microcosms in Ecological Research.* DOE CONF 781101.

Holling, C. S. 1986. Resilience of ecosystems: local surprise and global change. pp. 292–320. In Clark. W. C. and R. E. Munn (eds.). *Sustainable Development of the Biosphere.* Cambridge University Press. Cambridge, UK.

————. 1994. New science and new investments for a sustainable biosphere. pp. 57–73. In Jansson. A. M. M. Hammer. C. Folke. and R. Costanza (eds.). *Investing in Natural Capital.* Island Press, Washington. DC.

Hutchinson. G. E. 1948. Circular causal systems in ecology. *Annals New York Academy of Science* 50:221–246.

Jansson. A. M.. M. Hammer. C. Folke. and R. Constanza (eds.). 1994. *Investing in Natural Capital.* Island Press. Washington. DC.

Jansson. A.. and B. O. Jansson. 1994. *Ecosystem Properties as a Basis for Sustainability In Investing in Natural Capital.* Island Press. Washington. DC.

Kaplan, S. and B. J. Garrick. 1981. On the quantitative definition of risk. *Risk Analysis* 1:11–27.

King, D. M. 1994. Can we justify sustainability? New challenges facing ecological economics, pp. 323–342. In Jansson, A. M., M. Hammer, C. Folke, and R. Constanza (eds.), *Investing in Natural Capital.* Island Press, Washington, DC.

Kopp, R. J. and V. K. Smith (eds.). 1993. *Valuing Natural Assets—The Economics of Natural Resource Damage Assessments.* Resources for the Future, Washington, DC.

Levin, S. A. and K. D. Kimball (eds.). 1984. New perspectives in ecotoxicology. *Environmental Management* 8:375–442.

Lindeman, R. L. 1941. Seasonal food-cycle dynamics in a senescent lake. *American Midland Naturalist* 26:636–673.

Margalef, R. 1968. *Perspectives in Ecological Theory.* University of Chicago Press, Chicago, IL.

Mollison, B. 1990. *Permaculture. A Practical Guide for a Sustainable Future.* Island Press, Washington, DC.

Morgan, M. G. and M. Henrion. 1992. *Uncertainty: A Guide to Dealing with Uncertainty in Quantitative Risk and Policy Analysis.* Cambridge University Press, Cambridge, UK.

Niemi, G. J., N. E. Detenbeck, and J. A. Perry. 1993. Comparative analysis of variables to measure recovery rates in streams. *Environmental Toxicology and Chemistry* 2:1541–1547.

Odum, E. P. 1969. The strategy of ecosystem development. *Science* 164:262–270.

Odum, H. T. and R. C. Pinkerton. 1955. Time's speed regulator: The optimum efficiency for maximum power output in physical and biological systems. *American Scientist* 43:331–343.

——— . 1994. The energy of natural capital. pp. 200–214. In Jansson, A. M., M. Hammer, C. Folke, and R. Constanza (eds.), *Investing in Natural Capital.* Island Press, Washington, DC.

O'Neill, R. V., D. L. DeAngelis, J. B. Waide, and T. F. H. Allen. 1986. *A Hierarchical Concept of Ecosystems.* Princeton University Press, Princeton, NJ.

O'Neill, R. V., R. H. Gardner, L. W. Barnthouse, G. W. Suter, S. G. Hildebrand, and C. W. Gehrs. 1982. Ecosystem risk analysis: a new methodology. *Environmental Toxicology and Chemistry* 1:167–177.

O'Neill, R. V., S. M. Bartell, and R. H. Gardner. 1983. Patterns of toxicological effects in ecosystems: a modeling study. *Environmental Toxicology and Chemistry* 12:451–461.

Parkhurst, B. 1993. Framework for ecological risk assessment. In Bender, E. S. and F. A. Jones (eds.). *Applications of Ecological Risk Assessment to Hazardous Waste Site Remediation.* Water Environment Federation, Alexandria, VA.

Peet, J. 1992. *Energy and the Ecological Economics of Sustainability.* Island Press, Washington, DC.

Pickett, S. T. A. and P. S. White. 1985. *The Ecology of Natural Disturbance and Patch Dynamics.* Academic Press, Orlando, Florida.

Rango, A., J. Foster, and V. V. Salomonson. 1975. Extraction and utilization of space acquired physiographic data for water resources development. *Water Resources Bulletin* 11:1245–1255.

Scarano, L. J. and D. M. Woltering. 1993. Terrestrial and aquatic ecoassessment for a RCRA hazardous waste site, pp. 61–72. In Bender, E. S. and F. A. Jones (eds.). *Application of Ecological Risk Assessment to Hazardous Waste Site Remediation.* Water Environment Federation, Alexandria, VA.

Smith, F. E. 1975. Ecosystem and evolution. *Bulletin of the Ecological Society of America* 56:2–6.

Suter, G. W., II. 1989. Ecological endpoints. In Warren-Hicks, W., B. R. Parkhurst, and S. S. Baker, Jr. (eds.). *Ecological Assessment of Hazardous Waste Sites: A Field and Laboratory Reference Document.* US EPA 600/3-89/013, Washington, DC.

———. II. (ed.). 1992. *Ecological Risk Assessment.* Lewis Publishers, Chelsea, MI.

———. II and S. M. Bartell. 1992. Chapter 9. Ecosystem level effects, pp. 275–310. In Suter, G. W., II. (ed.). 1992. *Ecological Risk Assessment.* Lewis Publishers, Chelsea, MI.

Tansley, A. G. 1935. The use and abuse of vegetational concepts and terms. *Ecology* 16:284–307.

USEPA (U.S. Environmental Protection Agency). 1992. *Framework for Ecological Risk Assessment.* EPA/630/R-92/001. Washington, DC.

USEPA (U.S. Environmental Protection Agency). 1993. *A Review of Ecological Assessment Case Studies From a Risk Assessment Perspective.* EPA/630/R-92/005. Washington, DC.

Weinberg, G. 1975. *An Introduction to General Systems Thinking.* John Wiley, New York.

Whittaker, R. H. 1978a. *Classification of Plant Communities.* Dr. W. Junk Publishers, The Hague, Netherlands.

———. 1978b. *Ordination of Plant Communities.* Dr. W. Junk Publishers, The Hague, Netherlands.

CHAPTER FOUR

Estuarine Ecosystem Issues on the Chesapeake Bay

Walter R. Boynton

Introduction

On a global basis, estuarine systems constitute a small percentage (~0.5 percent) of the world's oceanic areas. However, the very high fisheries production (~21 percent of world's catch), proximity to major urban areas and transportation networks, and the use of these areas for recreational purposes make them far more important than indicated by spatial extent alone (Houde and Rutherford 1993). In part because of the location of these systems at the margin between land and ocean, serious degradation has become widespread during the last few decades. If current demographic projections are correct, we should expect that human activities in the coastal zone will continue to intensify. In 1988, for example, the average population density in coastal counties in the northeast region of the United States (Maine to Virginia) was about 340 people per square mile; it is expected to increase by an additional 30 percent by 2010. Sediments, nutrients, and an array of toxic materials will probably find their way into these aquatic systems, leading to further declines in water quality, habitat conditions and living resources, especially if these areas do not have effective management programs (Culliton et al. 1990). In addition, increased human activities will intensify pressures on the habitats and living resources characteristic of these systems. In many systems, seagrass communities and other habitats have already been lost or degraded, tidal wetlands filled, and fish and shellfish stocks overfished or contaminated. In many ways, rapid and poorly designed development and other activities within adjacent drainage basins have destroyed or negatively impacted the very resources which were the prime reasons stimulating development in the first place. A key question, which includes both economic and ecological concerns, is how to manage these systems for sustainable outputs of inextricably coupled economic and environmental products and characteristics.

In Maryland and Virginia, much attention has focused for several decades on the Chesapeake Bay and its tributary rivers. In the 1950s, descriptive scientific information was gathered, species identified, life history patterns clarified and advances made concerning the physics of the system. In the 1960s, some of these activities were continued and others added but, of relevance here, the first indications of water quality deterioration were noted but largely ignored. It was not until the early 1980s that a strong consensus developed as to the major problems facing the Bay environment, and in the late 1980s remedial management actions were developed (Malone et al. 1994). In addition, it was not until the 1980s that serious attention was paid to activities in the watersheds that discharge into estuarine systems such as Chesapeake Bay and, as such, are the sources of many of the problems confronting Bay ecosystems. There remains today considerable debate about the most prudent ways to manage activities in the drainage basins.

The overall purpose of this paper is to present information concerning contemporary ecosystem issues in Chesapeake Bay. To accomplish this goal, some information concerning important estuarine ecosystem characteristics is presented to familiarize those not from the environmental sciences with central issues. Patterns of change during the last several decades in selected ecosystem characteristics are also presented and the cause-effect linkages responsible for these changes described. Finally, management actions designed to improve the general health of the Bay are described.

Some General Organizing Principles
of Estuarine Ecology

Estuaries such as Chesapeake Bay are the ecosystems located on the margins that join continental lands with their surrounding seas. Estuarine ecosystems are coastal indentations
that have "restricted connection to the ocean and remain open at
least intermittently" (Day et al. 1989). In many of these ecosystems sea water is diluted by freshwater runoff from the land, but
in regions where evaporation is high or rainfall low, estuarine
salinities may be equal to or higher than those of the ocean. Most
present-day estuaries were formed during the last 15,000 years
of the current interglacial period, and are geologically recent features of the landscape (Day et al. 1989).

Because of the position of estuaries at the land-sea margin
throughout the world, there is considerable diversity in estuarine types. Recognizing these differences is important because
they influence the types of ecosystems that develop as well as the
susceptibility of these systems to impacts from human activities.
The most generally used classification of estuarine systems is
based on geomorphology (Pritchard 1952) and includes the following types: (a) lagoons or bar-built estuaries (e.g. the coastal
bays along the Maryland Atlantic coast), which are most often
oriented parallel to the coast, tend to be shallow (often less than
2 meters in depth), and generally lack vertical stratification of
the water column; (b) fjords, which result from glacial scouring,
are generally deep (>100 m) and "U" shaped in cross-section,
have strong vertical stratification and a sill or subsurface shoal-
ing at the seaward end, which limits exchange with the ocean; (c)
tectonically created estuaries (e.g. San Francisco Bay) which exhibit a variety of characteristics common to some of the other estuarine types; (d) coastal plain estuaries, which formed when river valleys became flooded after the last glaciation and have
moderate water column stratification, broad shoal areas, and a
moderately deep (~20–50 meters) central channel.

Chesapeake Bay is one of the best studied of the coastal plain
type estuarine systems. Many of the basic characteristics of this
system are well understood and have a direct bearing on ecological issues which are currently being confronted. One of the major, and perhaps unique, characteristics of the Chesapeake is the

very large drainage basin compared to either the surface area or volume of the receiving waters (Fig. 4–1).One primary effect of this is that there is very little potential for dilution of pollutants to harmless concentrations. In engineering parlance, "dilution is not the solution to pollution." at least in the case of the Chesapeake system.

A second feature of overriding importance concerns water movement patterns or circulation of Bay waters. Figure 4–2 presents a generalized schematic of bay circulation, where freshwa-

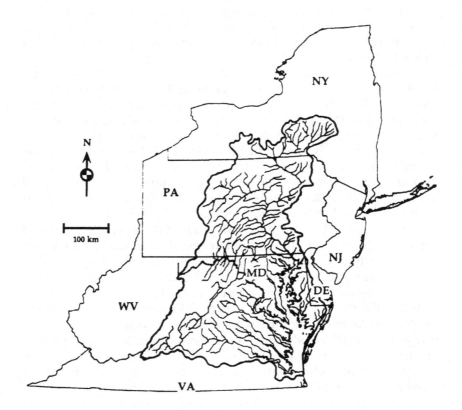

Figure 4–1. A regional view of Chesapeake Bay and its watershed showing the states encompassed and major portions of the tributary network. The ratio of drainage basin surface area to estuarine surface area is 28:1, indicating the potentially large impact of the land on this system. Inputs of water, nutrients, and organic matter are monitored at the fall-line of all the major rivers (representing 82 percent of the drainage basin); remaining loads are estimated using a land-use model.

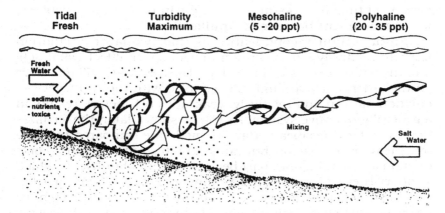

Figure 4–2. A simplified schematic diagram showing the main features of a two-layered estuarine circulation pattern. The major salinity zones along the estuarine salinity gradient are also shown. The bidirectional flow (seaward on the surface and landward on the bottom) acts to retain inputs which enter the system from the landward end. Numerous estuarine species have also adapted their life cycles to estuarine circulation patterns. For example, young blue crabs tend to stay in bottom waters when they are spawned near the ocean end and "ride" the up-estuary bottom water flow to the rich feeding grounds of the mesohaline and turbidity maximum regions. This diagram was adapted from Boicourt (1992).

ter from the drainage basin is shown entering on the left and moving toward the ocean as a surface water flow. To counter this seaward flow of freshwater, seawater moves into the bay as a near-bottom flow. This "gravitational circulation" is the net result of differences in pressure gradients which result from differences in the density of fresh and salt water (freshwater, being less dense than seawater, "floats" on top of the saltier bottom water). This bi-directional flow is characteristic of average conditions, but some degree of mixing occurs between the layers; mixing is more pronounced in some zones of the estuary (the turbidity maximum region) than in others (Fig. 4–2).

While there are many chemical and biological consequences due to this form of circulation, two are particularly important here. The first is that the vertical differences in density result in water column stratification, which in turn inhibits mixing of deep and surface waters. Despite the shallow nature of the Bay (mean depth ~10 meters), stratification is a very effective barrier, particularly from spring through early fall of most years. As a result, deeper waters are not exposed to the atmosphere for long periods of time (~weeks-months) and can become very depleted in oxygen, in part because of stratification.

An important second feature is that two-layer circulation leads to relatively long retention times . . . in effect, what gets into the bay tends to stay in the bay. The freshwater fill times are on the order of a year, and this, coupled with the two-layer circulation pattern and relatively weak tides (<1 meter), results in a generally retentive system. As will be shown later, the fertilization rates for Chesapeake Bay are moderate compared to many other estuarine systems, but rates of both plant and animal production are very high, in part because essential nutrients which support these processes are retained in the Bay rather than rapidly transported to the coastal ocean. Obviously, the retentive characteristic of the bay system is both a blessing (high ecosystem production rates) and a curse (retention of pollutants).

Finally, the bay ecosystem is characterized by very substantial temporal and spatial variabilities. Important inputs to the bay (i.e. freshwater, sediments and nutrients) vary strongly throughout the year (~10x> in spring versus fall) and vary between years (>2x) as well. These pulsing inputs and interannual variations in turn influence both plant and animal production and spatial distributions of these creatures. As a result of these variabilities, it has been and continues to be difficult to separate clearly the influence of such things as normal climatic variability from human-induced changes to these ecosystems resulting from pollutant inputs (see Brush, this volume).

Major Ecosystem Issues of Chesapeake Bay

For the last 15–20 years, there has been intense debate which first focused on whether there were ecological problems associated with Chesapeake Bay, and more recently on what those problems were specifically and what could be done to correct damaged portions of the ecosystem. These debates continue to this day relative to some old and emerging issues. In fact, the list of real or suspected ecosystem issues concerned with Chesapeake Bay is large and would be much larger still if the drainage basin of the Bay were to be included in this discussion. In the context of this paper, only a few can be discussed. The three ecosystem issues listed below were selected because there is general agreement that these are real problems, they have seriously impacted portions of the Bay ecosystem, and the cause-effect linkages are more or less understood, allowing for potential remedial management actions.

Over-fertilization, Algal Blooms, and Oxygen Depletions

There is clear evidence that fertilization of the Bay with nitrogen and phosphorus began to increase shortly after European settlement, due mainly to land clearing. Coupled with this, there were changes in Bay plant communities. In the last several decades these changes have accelerated, and now large algal blooms are common in some regions of the Bay. Due to the decomposition of these algal blooms, dissolved oxygen concentrations in deep waters of the Bay have become depleted to very low levels during the late spring-early fall period. Dissolved oxygen depletions have become more severe and covered larger regions of the Bay and some tributaries in recent years. Regions of the Bay with low (<2 mg l^{-1}) dissolved oxygen conditions represent habitats that are not available to any of the animals commonly associated with productive estuarine food webs (Boicourt 1992).

Seagrass Decline

Prior to the early 1960s, the shoal waters of Chesapeake Bay were dominated by a diversity of submersed vascular plants. In the decade of the 1970s, ten or more species were virtually eliminated from this estuarine environment. Submersed plant communities contributed significantly to food production for Bay fish, invertebrates, and waterfowl populations, to habitats used by small animals for refuge from predation, to stabilization of sediment processes, and to the rapid and efficient cycling of important chemical elements (Kemp et al. 1984).

Fishery Declines and Failures

Because of the obvious economic values associated with commercial and recreational fisheries, considerable attention has focused on the status of these organisms. During the late 1970s and 1980s, one species (*Morone saxatilus*, striped bass) underwent a severe stock decline and persistent recruitment failure, but recently responded favorably to strong management actions. Fishery yields of American oysters (*Crassostrea virginica*) have declined to record low levels, and remaining stocks are being further depleted by mortality associated with two diseases. Additionally, stocks of an historically important anadromous species, American shad (*Alosa sapidissima*), have been very depressed for several decades due to a number of factors including blockage (due to dams) of streams leading to spawning areas, intense harvest pressures, and acidification of spawning streams.

Changes in Drainage Basin and Estuarine Characteristics

Qualitative reports of Chesapeake Bay made during the seventeenth century through the middle of the present century clearly indicate that bay habitats and living resources, in the forms of fish, shellfish, and water fowl, were indeed abundant and played an important role in the economy of the region and the ecology of Chesapeake Bay. For example, William Penn in the late 1600s noted the extreme abundance of seafood as well as the huge size of oysters in the Bay, and in 1884 annual oyster harvests reached an historic peak of 20 million bushels. The writer H.L. Mencken noted in 1940 that "Baltimore lay very near the immense protein factory of Chesapeake Bay, and out of it ate divinely."

In the last few decades reports concerning various fisheries and habitats of the Bay have not been as positive, and there have often been calls for drastic action to rehabilitate the living resources of the Bay. In addition, the habitat diversity of the Bay appears to have been greater prior to the last few decades. Some 13 species of submersed aquatic vegetation ringed the Bay shores from the tidal fresh rivers to the high salinity waters near the mouth of the Bay; the water column was reasonably clear with sunlight penetrating to several meters in most areas, and deeper in the more saline portions, and was sufficient to support nutritious benthic algal communities; oyster reefs provided important topographical relief on the broad shoals of the Bay; cooler and deeper waters in the natural channels of the Bay provided a refuge from high summer temperatures for a variety of finfish. In recent decades a considerable fraction of these habitats has been lost. While large efforts are currently underway to restore Bay fisheries and habitats, there have been serious losses of both during the post World War II period.

Watershed Characteristics

Despite the fact that the Bay is embedded in a relatively huge watershed (Fig. 4–1) that has been continually modified by human activities, the watershed approach (as it is now called) was not part of the general scientific thinking or management actions until the 1980s. An analysis of the history of water quality studies in the Chesapeake found only a few calls for consideration of watershed impacts on the Bay prior to the 1980s, but serious action was not started until the initiation of the multi-state EPA Chesapeake Bay Program and the results of scientific studies that clearly tied discharges from the land to water quality and habitat conditions in the Bay. Since then much has been written concerning changes in the watershed of the bay (see USEPA 1992 for detailed treatment of watershed modifications).

Changes in the Bay ecosystem are not solely recent phenomena but date back to at least the early portions of the Colonial period. Using various chemical and biological markers in the sediments of the Bay, Cooper and Brush (1991) were able to show that: sedimentation rates in the bay increased 5 to 7 fold after 1760 due to land clearing; occurrences of anoxic conditions became more common after 1940; and the diversity of diatoms (an important unicellular plant) has decreased and changed to favor planktonic as opposed to benthic (sediment) forms, presumably because the Bay has become more turbid and less light penetrates to the bottom (see also Brush, this volume).

In more recent decades, both population and land use in the Chesapeake basin have continued to change. For example, at the close of World War II the population of the basin was just over 5 million; growth was especially rapid between 1960 and the mid-1970s, a period corresponding to important indications of ecosystem stress and change in Chesapeake Bay ecology (Fig. 4–3a). Associated with population growth, land use changes were also occurring. Specifically, forested lands increased, agricultural lands decreased (especially pasture lands), and urban and residential lands expanded (Fig. 4–3b). While forested lands tend to conserve nutrients and sediments quite effectively, other land uses export nutrients to a far greater extent. With contraction of agricultural lands came more intensive use of remaining

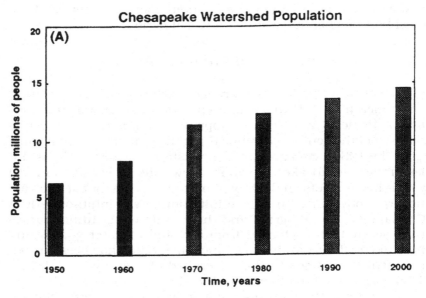

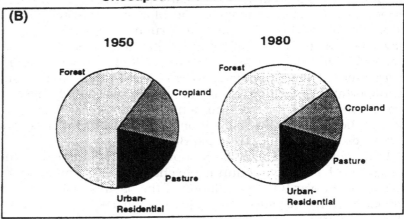

Figure 4–3. Summary of selected information concerning features of the Chesapeake drainage basin, including: (A) bar graph showing basin population from several different time periods (data are from USEPA 1992): (B) pie diagrams of land uses in the Chesapeake Bay drainage basin for two different time periods. Despite the increase in forested lands, nutrient loading rates for most portions of the Chesapeake system sharply increased in the period between 1950 and 1980. Since the late 1980s loads have decreased in some portions of the system and stabilized or increased only slowly in others (data are from USEPA 1983).

lands and of commercial fertilizers and pesticides, the use of which increased rapidly during the late 1960s and 1970s. The net effect of these changes in population and land use has been increasing loads of pollutants, especially nitrogen, phosphorus and sediments, to the Chesapeake Bay system.

Over-fertilization, Algal Blooms, and Oxygen Depletions

During the past few years, nutrient loading rates for a diverse mixture of ecosystems have appeared in the scientific literature, and it now seems safe to conclude that coastal systems have become among the most heavily fertilized of ecosystems because of increasing anthropogenic additions of nitrogen and phosphorus.

A few regions of Chesapeake Bay have been monitored for multiple decades, and in these areas it is possible to track changes in nutrient loading rates. In addition, it is possible to make crude estimates of what nutrient loading rates were at the time of initial European colonization of these systems and hence develop an estimate of pristine conditions, at least regarding nutrient loading rates (Table 4–1). In both the Patuxent and Potomac rivers, the historical record indicates increasing nitrogen inputs up to the present time; current loads exceed pristine loads by more than a factor of 5. Phosphorus loads also increased until the late 1970s and then decreased sharply in response to phosphorus removal at sewage treatment plants, a phosphate ban in detergents, and improved sediment erosion controls (phosphorus is readily transported via attachment to sediment particles). It is the goal of the Chesapeake Bay Program to reduce mid-1980s nutrient inputs by 40 percent by the year 2000.

Compared to other estuarine systems, nutrient loading rates to Chesapeake Bay are moderate to high for nitrogen and low to moderate for phosphorus. However, it is also clear that comparable nutrient loading rates in different ecosystems do not produce the same responses as those observed in the Bay. For example, nitrogen loading rates for the Potomac River and Narragansett Bay, Rhode Island are very similar, but poor water quality conditions extend throughout the mesohaline portion of the Potomac, whereas the analogous location is limited to a very restricted reach of upper Narragansett Bay (Magnien et al. 1990;

Table 4–1.   A comparison of estimates of annual nitrogen and phosphorus inputs to several well studied tributaries of the Chesapeake Bay system. Loading rates for the pre-european period (prior to 1600) were made by using nitrogen and phosphorus release rates from mature forests not exposed to significant atmospheric deposition of nitrogen and phosphorus; estimates for other periods were based primarily on direct measurements. Data in the table are from Boynton et al. (1995).

		Annual Nutrient Loading		
Location	Time Period	Total Nitrogen Load kg N × 10^6/yr	Total Phosphorus Load kg P × 10^6/yr	Reference
Patuxent River	Pre- European	0.37	0.01	Boynton et al. 1995
	1963	0.91	0.17	Jaworski 1992
	1969–71	1.11	0.25	Jaworski 1992
	1978	1.55	0.42	Jaworski 1992
	1985–86	1.73	0.21	Boynton et al. 1995
Potomac River	Pre- European	4.6	0.12	Boynton et al. 1995
	1913	18.6	0.91	Jaworski 1992
	1954	22.6	2.04	Jaworski 1992
	1969–71	25.2	5.38	Jaworski 1992
	1977–78	32.8	2.51	Jaworski 1992
	1985–86	32.1	3.35	Lugbill 1990
	1985–86	35.5	2.93	Boynton et al. 1995

Nixon et al. 1986). On the other hand, loading rates to the Baltic Sea are much lower than those of most of the Chesapeake systems, but hypoxic and anoxic conditions are now characteristic of both (Larsson et al. 1985). Estuarine morphology, circulation, and regional climate conditions undoubtedly have strong influences on the relative impact of nutrient loading rates (Wulff et al. 1990).

   In the Chesapeake Bay there is now strong evidence of the effects of increased nutrient fertilization, and in the last decade debate has been refocused from whether there were fertilization effects to how to achieve nutrient load reductions. One of the prime estuarine responses to nutrient fertilization is increased growth of phytoplankton, the unicellular plants that comprise the base of the food web. To a large extent, enhanced phytoplankton growth is analogous to the response of agricultural crops to fertilization. One of the most comprehensive evaluations of increased phytoplankton abundance in the Bay was developed by Harding (1994), who used both historical records of algal abundance and current aerial remote sensing data to develop a time series of observations covering four decades. In both the fresher and saltier regions of the Bay, there have been unmistakable increases in phytoplankton abundance which parallel increases in nutrient loading rates. Additional evidence for the linkage between nutrient loading rates and phytoplankton responses has been observed from data collected during the last decade of the Chesapeake Bay Monitoring Program; annual average phytoplankton abundance was strongly related to nutrient additions (Magnien et al. 1990).

The ecological effects of elevated phytoplankton abundance are of central concern. In the agricultural model, increased fertilization leads to larger crop yields and the overall effect is positive. However, in estuarine waters fertilization beyond a certain point initiates series of negative impacts, the results of which are propagated to varying degrees throughout the ecosystem. One of the initial effects occurs when abundant phytoplankton communities die and begin to decompose, mainly in the deeper waters of the Bay. In the process of decomposition, dissolved oxygen is consumed in large quantities and hypoxic (low oxygen) or anoxic (no oxygen) conditions result which are inhibitory or lethal to resident animal communities. It appears that the extent and duration of hypoxic conditions have increased since the 1950s (Cooper and Brush 1991). In addition, Boicourt (1992) has reported that the annual volume of hypoxic water in the Bay is a function of river flow; hypoxic volume increases as does river flow, due in part to the fertilization effect of the nutrients contained in river water and in part to the fact that the stratification of the Bay is proportional to river flow. In years of high flow, the resulting strong stratification prevents oxygen from the atmosphere from mixing into deep waters of the Bay and replenishing oxygen stocks depleted by decomposition of phytoplankton.

Seagrass Patterns

Submersed aquatic vegetation (SAV) communities play an important role in the functioning of shallow water portions of estuaries as well as in other aquatic ecosystems. Specifically, studies conducted over the last decade in estuarine systems indicate SAV communities maintain water clarity in shallow areas by binding sediments and baffling near-shore wave turbulence, modulate nutrient regimes by taking up nutrients in spring and holding these nutrients until fall, and enhance food-web production by supplying organic matter and habitat conducive for rapid growth of juvenile organisms. In much of Chesapeake Bay, SAV communities (which include some 13–15 species) started to undergo a serious decline during the 1960s in the upper portions of the Bay and in the early 1970s in the middle reaches of the Bay (Kemp et al. 1984). This decline was not taken seriously until the late 1970s, when a series of studies investigated potential causes. These experiments included field observations, small (50–700 liter) and large (400 cubic meters) microcosm exposure tests, and simulation modeling, and were conducted using several different plant species. Results indicated that the decline was primarily the result of nutrient over-enrichment. It appeared that epiphytic algae (a normal part of the SAV community) were over-stimulated by enhanced nutrient availability, which lead to increased shading of SAV leaves; photosynthetic rates of SAV were depressed below those needed for healthy plant growth. Increased water column turbidity and adhesion of suspended sediments to SAV leaves further reduced available light (Fig. 4–4). Herbicides were found to be a relatively small factor in the decline, although in areas of the Bay adjacent to agricultural drainage, seasonal herbicide stresses were possible (Kemp et al. 1984). It appears that if nutrient loading rates to these systems are reduced, SAV communities are capable of reestablishing themselves in many areas of the bay (Stevenson et al. 1993).

Selected Fishery Patterns

In the preceding portions of this paper, some of the more important changes in Chesapeake Bay ecology have been described; most have been negative. Before reaching the conclu-

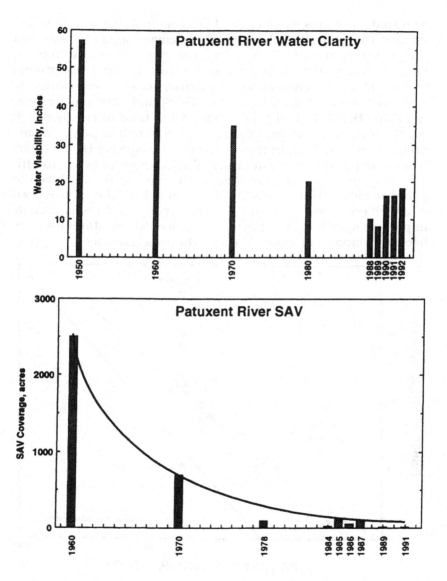

Figure 4–4. Historical patterns of water clarity and submerged aquatic vegetation (SAV) in the mesohaline zone of the Patuxent River estuary. Note that the rapid decline in SAV occurred during the decade of the 1960s, the same time period when the basin was undergoing large increases in population and changing land uses. Data are from U.S. Fish and Wildlife Service (1993).

sion that the Bay system is dead or close to it, we should be reminded that not all of the system is heavily impacted and that the degree of impact is worse in some years than in others. In fact, the Bay system is, as alluded to earlier, an immense protein factory. Data from a number of marine, coastal, and estuarine systems were organized by Nixon (1988) and explored for relationships between food production at the base of the food web (primary production) and fishery yields which depend on this production (Fig. 4–5). In this analysis. Chesapeake Bay is clearly a most productive system and furthermore appears to efficiently transfer organic material at the base of the food web to fisheries yields. Having said that not all is lost, there have been many changes in the status of commercially and recreationally important species in the Bay region since fishery statistics have been collected, and many of these changes have been negative.

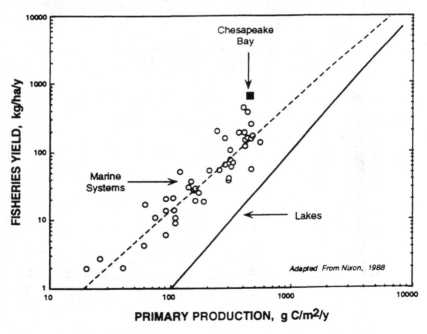

Figure 4–5. A scatter diagram showing relationship between primary production rates and fisheries yields for a variety of marine and estuarine ecosystems (open circles) and for a selection of lakes (summarized with the solid line). Chesapeake Bay is indicated by the bold square. The position of the Bay on the graph indicates that Bay food webs are particularly efficient in transfering food material into growth of commercial species. This figure was adapted from Nixon (1988).

The causes of these changes have been well established in some cases but remain unclear in others. In this section, an overview of recent trends of some of the most important fisheries of the bay is presented.

Striped Bass and American Shad The annual commercial catches of striped bass and American shad from both Virginia and Maryland and portions of Chesapeake Bay for the period 1929–1990 are shown in Figures 4–6a and 4–6c. In addition, an index of striped bass spawning success in Maryland waters is shown in Figure 4–6b. Striped bass catches underwent a multi-decade period of increase followed by a sharp decline in the last 15 years. At the present time, there is a relatively strict ban on striped bass fishing with only brief and highly regulated seasons in the spring and fall. Fishing for American shad is closed in the Bay and has been closed since the early 1980s, although fishing continues in waters of other coastal states. As with other commercially important species in the Bay, there is debate as to the most important causes of these declines. In the case of striped bass, reduction in spawning stock size (due to overfishing) and habitat degradation appear to be the most likely causes. The fishing ban on striped bass has yielded increased stock sizes, and 4–5 years after instituting the fishing ban, the regular pattern of successful recruitments every 2–4 years has started to reappear. There was a very successful striped bass recruitment in 1993 which is not shown in Figure 4–6b.

Despite the fact that the fishing ban on American shad has been in effect longer than the ban on striped bass, there is little indication that the stock has started to rebound, at least not to the degree observed for striped bass. One reason for this lack of response is that shad normally migrate much farther upstream than striped bass before spawning. Since virtually all tributary rivers of the Bay are dammed, it is generally assumed that blockage of access to optimal spawning areas has been a major factor impeding the re-establishment of this stock. In recent years fish ladders have been installed and operated in several important rivers, but the stock response has been small, suggesting that additional factors are also influencing recruitment. There are some data that suggest acid rain during the spring spawning season can depress pH in streams to levels lethal to shad larvae. In addition, the American shad fishing ban in Maryland does not apply to other coastal states; since shad spend a large percentage of their lives outside the Bay, they have been exposed to

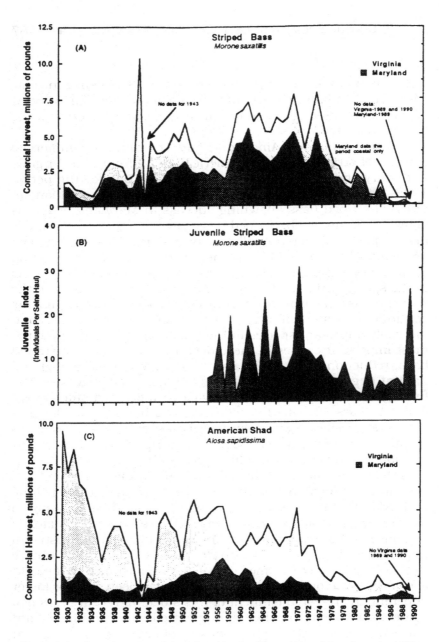

Figure 4–6. Annual commercial catches of striped bass (A) and American shad (C) from Maryland and Virginia waters between 1929 and 1990. Data are from Jones et al. (1990). Also shown is an index of striped bass spawning success (B) for the Maryland portion of the bay. Data are from Funderburk et al. (1991)

normal fishing pressure in these locations, and this may serve to inhibit stock rehabilitation.

American Oyster From 1929 through about 1960, combined Maryland and Virginia commercial oyster catches fluctuated between 20 and 40 million pounds per year. From 1960 through the early 1980s, combined catches decreased to 20–25 million pounds per year with virtually all of the decrease occurring in Virginia waters. However, there was a rapid decline in waters of both states beginning in 1981, and this trend has persisted and even intensified through the present time. There is considerable debate within the scientific, management, and fishing communities as to the relative importance of several factors in causing this decline, and there is equal if not more intense debate in management agencies concerning possible actions to rebuild this resource. Whatever management actions may eventually be taken, it appears that overfishing, disease, and loss of habitat have been the principle factors responsible for the decline of this resource.

Summary, Management Actions, and Ecosystem Responses

The major features of the ecosystem issues discussed above can be summarized in a simple cartoon diagram relating changing inputs of materials from the land to ecosystem outputs such as commercial fisheries (Fig. 4–7). In the diagram, nitrogen and phosphorus are shown entering the estuarine system and causing an increase in phytoplankton production and a decrease in light penetration due to shading by algae suspended in the water column. After nutrient supplies are exhausted by phytoplankton growth, the resultant blooms die and sink to the bottom, and oxygen is consumed in the decomposition of the bloom material. Hypoxic or anoxic conditions result, killing sessile organisms and removing the cooler deep waters as a habitat for fish communities. Nutrient enrichment also promotes the growth of algae on the leaves of SAV and limits the amount of light reaching the leaves; this light reduction, coupled with increased water turbidity, has been sufficient to kill SAV communities in many areas of the Bay. Again, SAV demise represents another loss of productive habitat in terms of a nursery and

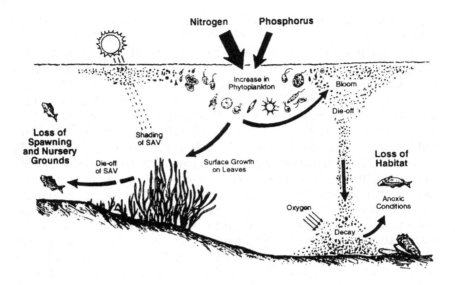

Figure 4–7. A cartoon diagram indicating the general cause-effect linkages of excessive nutrient (nitrogen and phosphorus) additions to estuarine ecosystems. When nutrient inputs become excessive, phytoplankton blooms occur and, after sinking to deep waters use large amounts of oxygen in decomposing. Excessive inputs also enhance algal growth on seagrass leaves leading to SAV die-off. Both the oxygen and SAV impacts represent serious habitat losses. This diagram was adapted from U.S. EPA (1983)

spawning area. Both of the habitat losses indicated in the diagram impact fisheries, but the quantitative relationships are not clearly established.

The Chesapeake Bay Program, which is a partnership between Bay states and the Federal government, has started on an ambitious program of bay restoration. The states are committed to reducing nutrient loads to the Bay by 40 percent by the year 2000; in several regions of the Bay, nutrient loads, particularly those associated with sewage treatment plant discharges, have already been reduced, and further reductions are planned. Most agree that control of diffuse sources of pollutants is the next big hurdle to be jumped and that this will be very difficult because of the dispersed nature of this nutrient source involving land use practices on private lands. Additional attention has been focused on nutrient additions from acid rain, which is a particularly important and only recently recognized source of pollution in the Bay region (Fisher and Oppenheimer 1991). Both the regional nature of the acid rain problem and the huge drainage basin of the Chesapeake make it clear that a regional, multistate approach to nutrient load reductions is necessary.

Studies have continued relative to SAV, and experimental planting programs indicate that the Bay grasses can be rehabilitated if nutrient loads are reduced (Stevenson et al. 1993). The experience agencies have had with the rehabilitation of the striped bass stock was particularly positive; a strong multi-state ban (or catch reductions) relieved fishing pressure to a point where the stock has started to successfully reproduce again. Management actions with other species have not been as successful, at least not to the present time. The disease problem with oysters appears to be particularly difficult to solve; this sector of the fishing industry is very depressed, and the path to rehabilitation is not clear. On a more positive note, there has been strong grassroots, state, and federal political support for improvements in the Bay environment, and because of this, it seems reasonable to expect that the search for solutions to Bay problems will continue and implementation of adequate pollution and fishing controls will lead to a healthier Chesapeake Bay.

References

Boicourt, W. C. 1992. Influences of circulation processes on dissolved oxygen in the Chesapeake Bay, p. 7–60. In D. E. Smith, M. Leffler and G. Mackiernan (eds.), *Oxygen Dynamics in the Chesapeake Bay: A Synthesis of Recent Research*, Maryland Sea Grant Program, College Park, MD.

Boynton, W. R., J. H. Garber, R. Summers and W. M. Kemp. 1995. Inputs, transformations and transport of nitrogen and phosphorus in Chesapeake Bay and selected tributaries. *Estuaries* (in press).

Cooper, S. R. and G. S. Brush. 1991. Long-term history of Chesapeake Bay anoxia. *Science* 254:992–996.

Culliton, T. J., M. A. Warren, T. R. Goodspeed, D. G. Remer, C. M. Blackwell, and J. J. McDonough III. 1990. *The Second Report of a Coastal Trends Series—50 Years of Population Change Along the Nation's Coasts. 1960–2010.* Strategic Assessment Branch, Ocean Assessments Division, Office of Oceanography and Marine Assessment, National Ocean Service, National Oceanic and Atmospheric Administration. Rockville, MD.

Day, J. W., Jr., C. A. S. Hall, W. M. Kemp and A. Yanez-Arancibia. 1989. *Estuarine Ecology.* John Wiley and Sons, New York, NY.

Fisher, D. and M. Oppenheimer. 1991. Atmospheric nitrogen deposition and the Chesapeake Bay estuary. *Ambio* 20(3):102–108.

Funderburk, S. L., J. A. Mihursky, S. J. Jordan, and D. Riley (eds). 1991. *Habitat Requirements for Chesapeake Bay Living Resources.* Chesapeake Research Consortium, Inc., Solomons, MD.

Harding, L. W., Jr. 1994. Long-term trends in the distribution of phytoplankton in Chesapeake Bay: roles of light, nutrients and streamflow. *Marine Ecology Progress Series* 104:267–191.

Houde, E. D. and E. S. Rutherford. 1993. Recent trends in estuarine fisheries: Predictions of fish production and yield. *Estuaries* 16(2):161–176.

Jaworski, N. A., P. M. Groffman, A. A. Keller and J. C. Prager. 1992. A watershed nitrogen and phosphorus balance: The upper Potomac River basin. *Estuaries* 15(1):83–95.

Jones, P. W., H. J. Speir, N. H. Butowski, R. O'Reilly, L. Gillingham and E. Smoller. 1990. *Chesapeake Bay Fisheries. Status, Trends, Priorities and Data Needs.* ASMF-MD-0048. Versar, Inc., Columbia, MD.

Kemp, W. M., W. R. Boynton, R. R. Twilley, J. C. Stevenson and J .C. Means. 1984. The decline of submerged vascular plants in upper Chesapeake Bay: Summary of results concerning possible causes. *Marine Techonology Society Journal* 17(2):78–89.

Larsson, U., R. Elmgren and F. Wulff. 1985. Eutrophication and the Baltic Sea: Causes and consequences. *Ambio* 14:9–14.

Lugbill, J. 1990. *Potomac River Basin Nutrient Inventory.* The Metropolitan Council of Governments, Washington, DC.

Magnien, R. E., D. K. Austin and B. D. Michael. 1990. *Chemical/Physical Properties Component.* Level I Data Report. Maryland Department of the Environment, Chesapeake Bay Water Quality Monitoring Program, Baltimore, MD.

Malone, T. C., W. Boynton, T. Horton and C. Stevenson. 1994. Nutrient loadings to surface waters: Chesapeake Bay case study, p. 8–38. In Uman (ed.), *Keeping Pace with Science and Engineering.* National Academy Press, Washington, DC.

Nixon, S. W. 1988. Physical energy inputs and the comparative ecology of lake and marine ecosystems. *Limnology and Oceanography* 33:1005–1025.

Nixon, S. W., C. D. Hunt and B. L. Nowicki. 1986. The retention of nutrients (C,N,P), heavy metals (Mn. Cd. Pb. Cu), and petroleum hydrocarbons in Narragansett Bay, p.99–122. In P. Lasserre and J. M. Martin (eds), *Biogeochemical Processes at the Land-Sea Boundary.* Elsevier Oceanography Series, 43. New York, NY.

Pritchard, D. W. 1952. Estuarine hydrography. In *Advances in Geophysics* 1:243–280. Academic Press, New York, NY.

Stevenson, J. C., L. W. Staver and K. W. Staver. 1993. Water quality associated with survival of submersed aquatic vegetation along an estuarine gradient. *Estuaries* 16(2):346–361.

U.S. Environmental Protection Agency, 1983. *Chesapeake Bay: A Profile of Environmental Change.* Chesapeake Bay Program, Annapolis, MD.

U.S. Environmental Protection Agency. 1992. *Chesapeake Bay Program. Progress Report of the Baywide Nutrient Reduction Reevaluation.* Annapolis, MD.

U.S. Fish and Wildlife Service. 1993. *Take Pride in Chesapeake Bay.* Annapolis, MD.

Wulff, F., A. Stigebrandt and L. Rahm. 1990. Nutrient dynamics of the Baltic. *Ambio* 14(3):126–133.

CHAPTER FIVE

Riparian and Terrestrial Issues in the Chesapeake: A Landscape Management Perspective

Curtis C. Bohlen

Rupert Friday

Introduction

The Chesapeake Bay watershed covers approximately 64,000 square miles, including portions of six states and the District of Columbia. This vast hydrologic network has been profoundly altered by a gradual transformation of the landscape that started with changes in forests caused by native American populations and accelerated soon after European settlement of the watershed 300 years ago.

The transformation of the watershed was accomplished without forethought or planning, by the cumulative effects of millions of decisions made by tens of thousands of landowners. Settlers cleared fields for homesteads, foresters harvested timber, builders constructed houses and towns, businesses built mills, warehouses and factories, farmers planted tobacco and small grains. Individuals, acting in response to what they thought best, within the institutional frameworks of their day, have transformed the land and transformed the Bay. When European settlers arrived, the Chesapeake watershed was more than 95 percent forest. Today the watershed is approximately 58 percent forest, 33 percent agricultural lands, 8 percent developed land—including low, medium, and high density residential, commercial, and industrial lands—and 1 percent water

(Neumiller et. al. 1994). Only about one half of the region's original wetlands remain (Tiner 1987). The decisions that led to these landscape changes were, and to a large extent continue to be, driven by human needs and wants on local spatial and short temporal scales, yet over time they have had profound effects at the scale of the Bay watershed.

Paleoecological information suggests that soon after the arrival of European settlers in the watershed, the ecology of the Bay began to change (Brush, this volume). Sedimentary records suggest that sedimentation rates climbed as forests were cleared for agriculture. Anoxic conditions in the Bay became more frequent, and signs of nutrient enrichment appeared in Bay sediments and ecosystems (Cooper and Brush 1991). A profound hydrologic alteration of the Bay also occurred. Increases in surface runoff and decreases in evapotranspiration throughout the watershed, triggered by the removal of forests, caused an increase in freshwater flows to the Bay, reducing salinity in the upper Bay. The State of Maryland now mines oyster shell from once-productive oyster bars in the upper Chesapeake, where salinity in the water is now too low to support oysters, and places the shell in the lower Bay, where oysters can still survive.

The transformation of landscapes within the Chesapeake watershed not only continues, but has accelerated. The population of the watershed, slightly over 13 million people today, is expected to increase by about 20 percent in the next quarter century. But development patterns have changed dramatically in the last 50 years, and consumption of land will climb faster than population. New development consumes nearly twice as much land per capita as existing development has (2020 Panel 1988). Thus, while forest loss in the Chesapeake watershed has slowed, and even reversed in some regions, urbanization and suburbanization have increased.

The potential implications of these trends for the Chesapeake and its tributaries are troubling. Without implementation of more sophisticated approaches to understanding and managing watershed-scale consequences of local land use decisions, continued loss of forest and wetlands, increased human populations, and more abundant roads, rooftops, parking areas and other "impervious surfaces" will increase flow of pollutants, especially nutrients, to the Chesapeake, and degrade both terrestrial and aquatic habitats. A variety of local environmental services, from provision of habitat for migratory birds and protection of human populations from flooding, are likely to be disrupted. And regional environmental services such as support of biodiversity, production of anadromous fishes and support of commercial fisheries will be increasingly strained. These reductions in environmental services represent real social costs of landscape change in the Chesapeake watershed, costs that in many cases could be reduced by consideration of landscape dynamics to guide policy and steer investments in environmental restoration and enhancement.

Landscapes, Scale, and Land Management

Landscapes are hierarchically structured (O'Neill et al. 1986, Forman and Godron 1986). Larger landscapes (e.g. the Chesapeake Watershed) are composed of smaller landscapes (counties, sub-watersheds), which, in turn, are composed of smaller units. The hierarchical nature of landscapes implies a dependence of dynamics (patterns of change over time) at one spatio-temporal scale on those occurring at other scales. Local changes and changes at the landscape scale are necessarily linked, if only because landscapes are built up of local-scale features. Those linkages, however, take particular forms because differences in the characteristic frequencies or response times at successive levels in hierarchical systems partially insulate each level from adjacent (higher or lower) levels (O'Neill et al. 1986; Holling 1992).

Phenomena occurring on landscape scales provide a slowly changing background for events at smaller, local scales (O'Neill et al. 1986). Thus, the Chesapeake Bay watershed provides a gradually changing context for phenomena occurring on individual land parcels. A decision to build a seafood processing facility or a commercial fishing pier, for example, is predicated on an abundance of fish, crabs, and oysters. As the Bay's production of these resources has declined, commercial coastal lands have become available for marinas, vacation homes, and other uses not as dependent on abundant seafood. The landscape dynamics are slow enough, however, so that local events—whether land use decisions or changes in the abundance of muskrats—are predominately controlled by the current condition of the Bay, and only secondarily by how the condition of the Bay is changing.

Similarly, rapid changes at local scales are often attenuated by the slow response times of landscapes. For example, the annual decisions farmers make selecting among commodities to produce on their farms induce short-term fluctuations in land cover and land use. These fluctuations, however, have only limited effects on the Chesapeake watershed as a whole. The watershed is too large and changes too slowly to respond to such short-term, local fluctuations.

The cumulative impacts of local land use decisions on watershed or landscape processes can be profound (Preston and Bedford 1988). The relationship between the health of a landscape and the health of component ecosystems, lands, and habitats, however, is a complex one. Both the *scale* of changes in land condition and the *location* of lands so affected are important for determining overall landscape response. A limited degree of agricultural or suburban development is possible within larger landscape units without serious impairment of landscape processes (Klein 1979; Schueler and Galli 1991). However, there are limits to this flexibility. When dynamically important lands within a landscape (e.g., wetlands, riparian areas, floodplains) are disturbed, or when ecological processes are altered or disrupted on a sufficient proportion of less sensitive lands, landscape-level processes and thus landscape-level environmental services may be impaired.

Current efforts to institute "ecosystem management" are, in part, efforts to recognize landscape-scale ecological and social processes that have traditionally been outside the range of consideration of land managers (Grumbine 1994, Lackey 1994).

Scale and the Management of Landscapes

Land managers operating at different spatial scales perceive different incentives for management action, and are capable of effectively managing different resources (Bohlen and King 1995). Land owners' primary management focus tends to be on-site resources. Local governments perceive the effects of development decisions on the local landscape, including effects on tax revenues, human health, costs of county services, aesthetics, and local environmental effects. Federal managers are charged with protecting resources at national scales, and attend to interstate resources, such as migratory birds and major rivers, that local and state governments are unable or unwilling to manage effectively.

The scale of environmental management necessary for supporting or enhancing environmental benefits depends on the particular benefit under consideration. Landowners, for example, are capable of effective management for timber, because most management actions to increase timber production can be carried out without reference to practices on adjacent lands. Since landowners also receive many of the benefits of managing their lands for timber (see Figure 5-1), investments in timber management are freely undertaken (provided harvest is not too far in the future). In contrast, managing a wetland or a stream reach to support stocks of anadromous fishes is impossible for most landowners. Even a high quality stream reach or wetland in a watershed that provides poor habitat for anadromous fishes will support few fish. In addition, many of the benefits of efforts to support anadromous fishes will accrue many tens of kilometers from the stream reaches in which the fish reproduce. Therefore land managers (like many landowners and local governments) who focus on local environmental benefits would have difficulty protecting anadromous fishes, but equally important, they may perceive weak incentives to do so, since they or their constituencies would receive only a small share of the benefits. Larger scale (regional or national) managers, on the other hand, serve larger constituencies that include those who benefit from improved commercial and recreational fishing downstream. Thus regional or national authorities are more likely to invest in protecting them.

The scale on which resources need to be managed does not always match those on which the benefits are most readily received, leading to resource conflicts. Resources that fall above the diagonal line in Figure 5-1 are hard for individual land owners to manage, even if they want to, because local benefits are supported by ecological processes occurring outside the local area. In the presence of institutions such as elected governments, markets, and voluntary associations that are capable of aggregating preferences of scattered individuals, cooperative, minimally coercive management may be possible (e.g., bag limits on waterfowl). Management may be minimally coercive in the sense that many local decision makers will receive benefits from management actions, and coercion is needed primarily to discourage "free riders."

Resources below the diagonal line, on the other hand, present fundamental conflicts of interest between landowners and others in society. The resources generally must be managed at small scales (often at cost to landowners), but they produce benefits that accrue primarily to others. Thus landowner-dominated decisions generate externalities, and there may be calls for

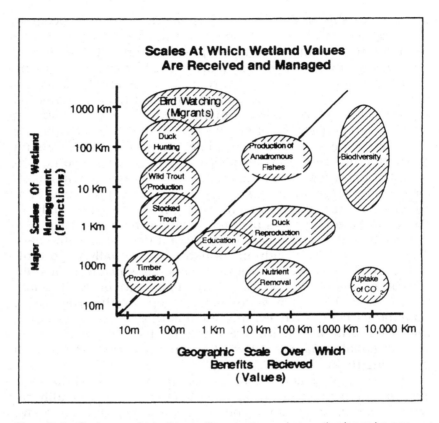

Figure 5–1. Scale over which site-specific resources such as wetlands can be managed do not always match the scales over which resulting benefits are realized. Externalities beyond the control of any single decision maker are common (modified from Bohlen and King 1995).

regulation in order to protect others' interests. Resulting regulations are coercive in the sense that they impose real costs on local decision makers, thus engendering non-compliance and political opposition. This conflict between local and regional benefits underlies much of the current political controversy over regulation and property rights.

Landscapes in the Chesapeake Region

In this paper we focus on two contrasting landscapes in the Chesapeake watershed. The two landscapes are (1) rapidly suburbanizing Anne Arundel County, Maryland, and (2) the largely agricultural Nanticoke river watershed of Maryland and Delaware. Neither landscape can be considered "natural." Both are products of a long history of human management. The Nanticoke watershed is highly agricultural. Anne Arundel county is in the midst of a suburban transition.

Consequences of Suburbanization

Forest clearing and the transformation of agricultural land into lawn, roads, and buildings trigger profound hydrological, physical, and chemical changes at the landscape scale.

Abundant impervious surfaces in urban and suburban landscapes (roads, parking lots and roofs) prevent water from infiltrating into the soil. Infiltration on what pervious areas remain is reduced in comparison to that which occurs in forested or even most agricultural landscapes. The little water falling on suburban landscapes finds its way into the ground water, where water flow rates are slow and opportunities for biological and physical removal of pollutants great (See Figure 5-3, page 104, Schueler 1987).

Under pre-development conditions, a substantial portion of precipitation enters the groundwater, which slowly drains to streams over weeks or months. Flow paths to surface drainage networks are long. Natural ephemeral and low order streams dissipate a substantial proportion of the energy of falling water in turbulent flow and friction, slowing water movement. Thus water levels in streams remain higher between precipitation events, maintaining sufficient base-flows to protect aquatic organisms, and the pulse of water that reaches the stream after a storm event (arrow in the figure) arrives slowly, spread out in time, resulting in a relatively low peak discharge.

Developed landscapes, on the other hand, are "flashy." Structurally complex natural drainage networks are replaced with simpler storm drain systems in which turbulent flow and friction are reduced. Water moves via engineered surface water conveyances, instead of via a combination of groundwater, surface sheet flow, and natural channels. Flood pulses are rapid and high, but between-storm flows are low; most of the stream flow occurs in brief flashes immediately following rainfall events.

The ecological and geomorphic consequences are substantial. First, flashy streams are erosive streams. High discharges and associated high water velocities in streams move soils and sediments much more effectively than do low velocity flows. The resulting stream bed erosion in suburban and urban streams in Anne Arundel County is sometimes severe enough to turn streams into biologically depauperate gullies. In one extreme case (the west branch of Weams Creek on the outskirts of An-

---

Case 1: Anne Arundel County

Anne Arundel County, Maryland is a suburbanizing county located east of Washington, D.C. and south of Baltimore. North of Annapolis, the county is highly suburbanized, while "south county" retains much of its rural character. Most of the county is now within an hour's drive of either Washington, D.C. or Baltimore, making the entire county attractive for suburban development. Deep, poorly consolidated coastal deposits underlie the county. Soils tend to be sandy and easily eroded. The Patuxent River, which forms the western boundary of the county, is the only large non-tidal river in the county. Most streams are small, and drain to tidewater within a few miles of their headwaters, either to the tidal Patuxent or to one of several tidal rivers to the east. Stream valleys are often steep-sided, with narrow but well developed riparian areas. Extensive forests remain in the central and western part of the county, with a few large forests in the north and abundant small woodlands to the south.

Land use changes in the county over the last 20 years have been profound (See Figure 5–2, page 104). Between 1973 and 1990, total developed land in the county increased from 28 percent to 35 percent of county land area. Low density residential development (which increased by almost 50 percent over the period) accounted for almost half of that increase. Over the same period, forest and agricultural lands declined by 10 percent. Almost half of the loss of forest and agricultural lands occurred in the last five years of the period, from 1985 to 1990 (all statistics on land use from Maryland Office of Planning 1991).

(continued)

> *(Continued from previous page)*
>   These statistics are a symptom of a general accelera-
> tion of land consumption in Anne Arundel County, in Mary-
> land as a whole, and in the Chesapeake Bay watershed
> (Maryland Office of Planning 1993, 2020 Panel 1988). The
> population of the state, while growing overall, has been
> abandoning developed areas and moving into newly subur-
> banizing areas. From 1970 to 1990, Maryland's urban and
> inner suburban areas have declined in population. The
> (mostly urban) areas of the state that declined in popula-
> tion over that 20-year period witnessed a 21 percent loss in
> total population, despite state-wide population growth. In
> contrast, newly suburban areas of the state have shown
> strong population increases. Existing suburban areas and
> most rural areas have had slight population increases
> (Maryland Office of Planning 1993). Per capita land con-
> sumption has increased substantially in the last few
> decades. As of 1950, an average of 0.18 acres of land had
> been developed per person in the Chesapeake Bay water-
> shed. By 1980, land intensity of development had increased
> to the extent that 0.65 acres of land were being developed
> per new Maryland resident (2020 Panel 1988).

napolis), the stream has cut a gully over 3 meters deep in places
(personal observation). Sediments from downcutting are also
transported downstream, where they go on to harm other aquatic ecosystems and trigger dredging and other activities to ameliorate their effects on waterfront land owners.

Second, urban and suburban streams are often unable to
support normal riparian communities. Riparian areas and associated wetlands trap sediments and remove nutrients, thus protecting downstream water quality. Riparian areas are also important to a variety of landscape-scale processes, including creation
of high-quality aquatic habitats, water storage and support of
biodiversity (Peterjohn and Correll 1984; Welsch 1991; Schlosser
1991; Richardson 1994; Bohlen 1992–93; Lowrance et al. 1995).
Downcutting of streams can dry out adjacent riparian areas and
wetlands, reducing water quality and other values. Even where
downcutting has not been severe, flashy conditions and reduced
infiltration shrink the residence time of waters in riparian areas,
lessening opportunities for biological processing of nutrients.

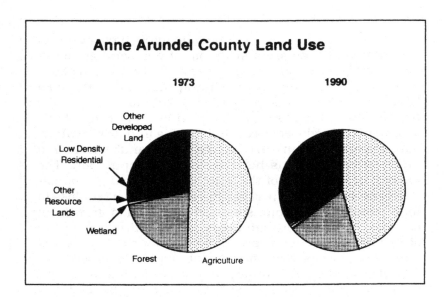

Figure 5–2.   Land use in Anne Arundel County. Maryland, 1973 and 1990. Agricultural and forest lands have declined, and low density residential lands and other developed lands have increased (based on data from The Maryland Office of Planning).

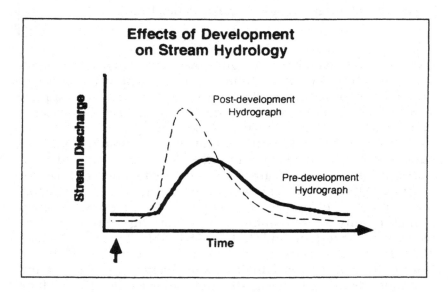

Figure 5–3.   Idealized Pre- And Post-Development Hydrograph For Maryland Streams (modified from Schueler 1987).

Third, urban and suburban streams provide poor habitat for most aquatic organisms. Abundance, number of species, and total diversity of stream fish and invertebrates generally decline with watershed imperviousness, which is highly correlated with the abundance of urban and suburban lands within the watershed (Klein 1979; Schueler and Galli 1991). With even moderate levels of suburban development and extensive use of urban Best Management Practices (BMPs), few species of fish survive, and those that do survive are generally of little recreational value. The biotic integrity of urban and suburban streams is generally low (Karr et al. 1985, Hall et al. 1994).

In addition to its physical effects, suburbanization also increases the flow of various pollutants to receiving waters. Urban and suburban landscapes release substantial quantities of nutrients, sediments, hydrocarbons, metals, and other pollutants into surface waters (EPA 1991; Schueler 1987; Ailstock and Horner 1991; Olsenholler 1991). Flows of nutrients have been especially problematic in the larger context of the Chesapeake restoration effort. In 1987, the Chesapeake Bay Program adopted an ambitious goal of reducing nitrogen and phosphorous flows to the Bay by 40 percent by the year 2000 (Chesapeake Bay Program 1994). Suburbanization increases nutrient flows in a number of ways:

1. Nitrogen and phosphorous are released from suburban landscapes simply because of their large human populations. Large quantities of nutrients are imported into the Chesapeake watershed in food. Those nutrients are seldom exported from the region, but are released into ground and surface waters via septic tanks and sewage treatment plants.

2. Maryland now has more acres in lawn than in corn production (Horton and Eichbaum 1990). Grass for ornamental purposes is thus one of the state's major "crops." Many lawns receive high levels of fertilizer, and leaching of nutrients can be significant.

3. Approximately a quarter of the total nitrogen entering the Chesapeake Bay is derived from atmospheric sources (MDE 1992). Of that, about one third is thought to be derived from automobiles (MDE 1993). In suburban landscapes, people are widely scattered and residences are far from shopping and work. Although emissions per vehicle mile traveled have fallen, increased

travel (in part encouraged by suburban development patterns) has more than made up for the difference.
4. Loss of forest also contributes to increased nutrient loadings since forests are both the region's least polluting land use and highly conservative of nutrients. Suburbanization replaces a non-polluting land use with a much more polluting one.

Sediment releases from suburban landscapes and from development sites also cause problems for aquatic ecosystems. Sediments fill navigation channels, reduce light penetration in the water column, bury benthic communities, reduce feeding efficiency of suspension feeders, and cause physical damage to gills and other delicate biological structures. In streams and rivers, sediments may also alter water flow patterns, sediment transfer processes, and stream bottom properties in ways harmful to fish and other desirable aquatic life. The impacts of sediments may be further exacerbated because phosphorus and a variety of toxic chemicals often travel adsorbed to sediment particles.

Consequences of Agriculture

Much of modern agricultural practice is an effort to keep agricultural fields in early successional states. Repeated disturbances (in the form of plowing, disking, cultivating, applying herbicides, and harvesting) prevent the development of later successional ecosystems (e.g., forests) that would be undesirable for agricultural production. Early successional ecosystems generally show an excess of primary productivity over respiration. In many agricultural production systems, that excess is captured for human use in the form of crops (Odum 1969). Unfortunately, early successional ecosystems also tend to be leaky. Early successional or disturbed ecosystems retain nutrients and sediments less tightly than do less frequently disturbed communities (Odum 1969; Bormann and Likens 1979). Thus agricultural efforts to maintain early successional conditions in order to produce crops are associated with releases of sediments and nutrients to adjacent ecosystems. Careful agricultural management efforts can limit the losses, but they are extremely difficult to eliminate entirely.

On a landscape scale, agriculture increases nutrient flows to the Bay by importing nutrients into the Chesapeake watershed,

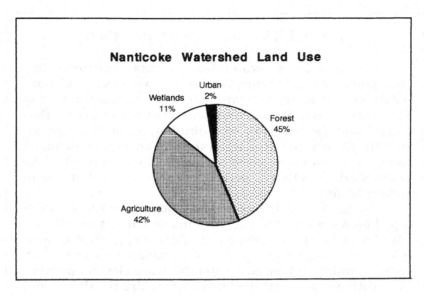

Figure 5–4. Land use in the Nanticoke watershed. Forested wetlands have been included in the "Forest" land use category. Total wetland area therefore is somewhat higher than this diagram suggests.

especially in the form of fertilizers and animal feeds. Nutrients are also exported from the region in agricultural products, but because of nutrient losses within the agricultural production system, a portion of the imported nutrients remains within the watershed, increasing total loadings to the Bay.

The agricultural community has, for years, worked hard to reduce soil erosion and the export of nutrients from agricultural lands. Various management practices, from fertilization practices, to strip cropping, to low-till or no-till farming systems, can increase or decrease the loss of nutrients and soil from agriculture. A variety of federal and state agencies have worked with farmers to develop management techniques to reduce the loss of sediment and nutrients from farm fields. Maryland now requires many farms within the 1000-foot "critical area" adjoining the Chesapeake and its tidal tributaries to have detailed nutrient management plans. These plans, which vary in their complexity, represent a concerted effort on the part of farmers to reduce the loss of nutrients from agriculture. The Soil Conservation Service worked for decades with farmers to implement agricultural Best Management Practices (BMPs) targeted at reducing soil loss. The

Case 2: The Nanticoke Watershed

The Nanticoke watershed drains approximately 400,000 acres in Caroline, Dorchester, Wicomico, and Somerset counties in Maryland, as well as approximately 315,000 acres in Sussex and Kent counties, Delaware (Nature Conservancy 1994). Agriculture accounts for approximately 42 percent of the watershed by area. Forests, many intensively managed, cover an additional 45 percent of the watershed. Less than 2 percent of the watershed area is in urban lands.

The landscape is one of low topographic relief, developed on a variety of unconsolidated sediments, mostly derived from sandy and silty coastal plain deposits. With little elevation change, water potential gradients are low, so both ground water and surface water flows are slow. In particular, without artificial drainage, water drains slowly and ponds extensively. Extensive wetland complexes were once found along drainage divides throughout the region where topographic gradients are low and drainage patterns ill-defined. In the upper, Delaware portion of the watershed, substantial areas were drained for agriculture, many of them as WPA projects during the Depression.

There has been little change in land use in the watershed over the last few decades, and only a slight increase in urban land over this period, although development of suburban strips along major roads has occurred around the watershed's larger towns and cities. Data from the Yearbooks of Agriculture (U.S. Department of Commerce 1954, 1959, 1964, 1969, 1974, 1982, 1987, and 1992) for Dorchester County, Maryland and Sussex County, Delaware show that between 1948 and the present, total cropland area has remained more or less constant, or increased slightly. Simultaneously, the total number of farms has declined, and average farm size has risen (Figure 5–5).

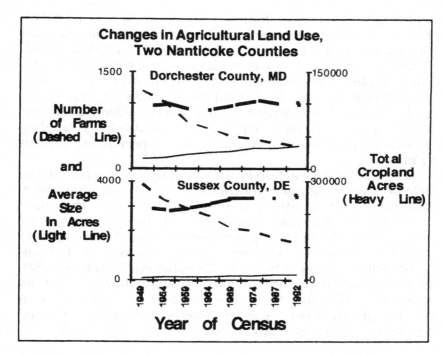

Figure 5–5. Changes in land use in two Nanticoke counties, 1948–1992. (Source: U.S. Department of Commerce).

SCS's mission was gradually broadened by legislation and policy to incorporate an increasingly wider range of resource protection issues, including protection of water quality, the reduction of nutrient runoff, wildlife conservation, and wetland protection. Near the end of 1994, the agency's name was officially changed to "Natural Resources Conservation Service" in order to reflect its broader mission.

Despite these and related efforts, agriculture remains a major source of nutrients to surface and ground water. Water quality in the tidal fresh portion of the Nanticoke river reflects the impacts of agriculture. The consequences are shown in water quality monitoring records collected between 1985 and 1989 at the tributary water quality monitoring station on the tidal freshwater portion of the Nanticoke river, near Sharpstown, Maryland (MDE 1994). Dissolved inorganic nitrogen (sum of nitrate, nitrite, and ammonium concentrations) was routinely available in excess, and sometimes in tremendous excess, of levels limiting to phytoplankton growth. Ratios of the concentrations of total nitrogen to total phosphorus were also very high, indicating that phytoplankton growth was generally limited by available phosphorous. Chlorophyll A concentrations (a measure of phytoplankton abundance) were high, and Secchi depths (a measure of water clarity) were low. Conditions were poor enough to preclude growth of submerged aquatic plants, which grow best when waters are clear. Ecologically, the Nanticoke is suffering negative effects from nutrient enrichment, especially enrichment by nitrogen.

The U.S. Geological Survey has found that agriculture on the Delmarva peninsula also contributes nitrate to ground water. Detectable levels of nitrate were found throughout the upper portions of the Nanticoke watershed, but concentrations of nitrate exceeded EPA drinking water standards (10 mg/l, approximately 15 times the level at which phytoplankton growth would be limited) only in a few hotspots (Hamilton and Shedlock 1992). Over time (sometimes measured in decades in the flat Nanticoke watershed), this nitrate-enriched ground water will flow to streams and other surface water systems, eventually increasing nitrate loadings to the Chesapeake.

Consequences of Drainage

The construction of drainage systems throughout the upper portion of the Nanticoke represents a significant change in landscape dynamics, and has had a series of ecological consequences.

Construction of drainage ditches successfully sped the removal of water from the landscape, thus allowing expansion of agricultural production. The same increase in the speed with which water drains off of the landscape, however, changes the patterns of water flow entering downstream ecosystems. Construction of drainage systems removes surface water, and in the sandy soils that predominate in much of the Nanticoke watershed can lower the water table aquifer as well. The net effect is to reduce the ability of landscape to trap and store water. Rainfall from storm events, instead of being stored within the upstream portions of the watershed, is now passed downstream, increasing peak discharge following storms, and increasing the variability of the salinity and other chemical characteristics of estuarine waters downstream.

Drainage has resulted in the loss of substantial areas of wetland and in the degradation of riparian areas. Wetlands and riparian areas can be remarkably effective at removing, transforming, or neutralizing sediments, nutrients, and certain other common agricultural pollutants from surface and groundwater flows (e.g., Moshiri 1993; Peterjohn and Correll 1984; Welsch 1991; Lowrance et al. 1995). The potential water quality benefits of these areas have been reduced because (1) the area of wetland and riparian forest within the landscape has been reduced, (2) ground and surface waters draining agricultural areas and carrying heavy loads of pollutants are now more likely to bypass riparian areas, and (3) those waters that do pass through a riparian area are likely to flow through the biologically active zones of the soil more quickly under the increased hydraulic gradient provided by drainage systems. These changes reduce the extent of denitrification, physical trapping of sediments, and biological uptake of nutrients within the upper portions of the watershed.

It is possible, although not yet proven, that hydrologic changes in the river have exacerbated the impacts of acid deposition on rockfish (*Morone saxitalis*). Rockfish larvae are unable to survive in the Nanticoke river, apparently because of low pH and elevated levels of aluminum in the water (Hall et al. 1985). The shorter time that water from storm events resides on the landscape reduces the extent to which storm waters are mixed with less acidic groundwater. Moreover, decreased contact of storm waters with wetland and riparian systems may limit biogenic buffering processes capable of reducing the detrimental effects of the acid deposition. Thus hydrologic modifications may have increased the severity of acidic "flashes" that follow precipitation events, indirectly limiting rockfish recruitment and reducing populations of other acid-sensitive fish species (Hall et al. 1994).

Landscape Management

Management of Landscapes Via Natural Vs. Cultural Processes

Landscapes maintain and change their physical structure through endogenous (ecological and physical) processes, through maintenance processes carried out by humans, or through some combination of the two. In human-dominated ecosystems, maintenance processes may derive from government expenditures, general economic activity, engineering, or other human behaviors. In unmanaged landscapes, ecological maintenance processes predominate. These non-anthropogenic maintenance processes are known as "functions" in the wetlands science and policy literature (Richardson 1994). We use the term "ecological processes" to emphasize that these phenomena need not have any utilitarian benefit to be significant from the perspective of landscape dynamics.

The choices land managers face can be depicted schematically in a Landscape Management Ternary Diagram (Figure 5–4). Human activity may sever or alter one or several of the environmental processes that maintain the ecological and physical structure of an unmanaged landscape. The resulting landscape change may be perceived either as beneficial (e.g., when agriculture is established) or as detrimental (e.g., when suburbanization leads to degradation of stream ecosystems). When humans like the landscape changes, cultural processes develop or are established that maintain the landscape in its new, desirable form. Therefore, the landscape moves from the lower right region of the ternary diagram labeled "Maintained by Ecosystem Processes" toward the top of the diagram, labeled "Maintained By Cultural Processes" to reflect the increased importance of human activity in maintaining the structure of the landscape. If humans do not like the landscape changes, two outcomes are possible. First, cultural management may not develop, and the landscape moves toward the lower left part of the diagram, labeled "Low Environmental Quality." Second, human societies may undertake defensive expenditures to resist the landscape changes, again increasing the degree to which the landscape is maintained by cultural processes, and moving toward the upper portion of the ternary diagram.

The Landscape Management Ternary Diagram depicts, in a schematic way, the relative intensity or importance of cultural and ecological processes in structuring landscapes. A healthy urban landscape, for example, might rest somewhere near the "1" on the diagram; a wilderness area, somewhere near the "2." An agricultural landscape dominated by conventional agricultural practices might lie at "3," while a small, heavily managed nature reserve in a suburban landscape might lie somewhere near the "4." More generally, the diagram represents a state space over which landscapes evolve as they are disturbed, managed, and abandoned by humans. The diagram provides a framework for understanding a variety of land management decisions.

It is important to point out that both high quality and low quality lands may be maintained predominately by either cultural processes or ecosystem processes. The relative importance of cultural and ecosystem maintenance is not, in general, related to any subjective evaluation of environmental quality. Relatively stable, culturally dominated landscapes—of both low and high quality—have existed in Europe and Asia for a millennium or more. Moreover, the conceptual separation we are using between cultural and ecosystem processes is in no way an effort to separate humans from nature. In fact, it is in large part motivated by an effort to develop analytic alternatives to a simplistic dichotomy between natural landscapes and other (unnatural?)

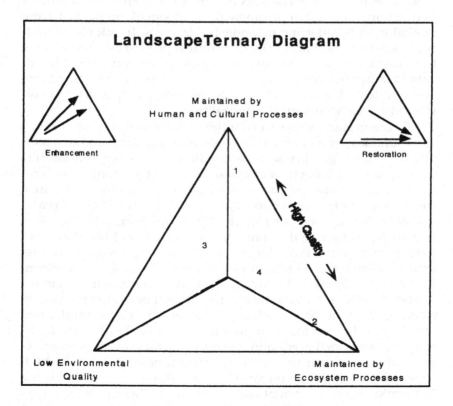

Figure 5–6. The Landscape Management Ternary Diagram.

landscapes. If such a dichotomy ever had any validity, it certainly has little in a world in which even the global radiation budget has been altered by human activity (McKibben 1989). Essentially all landscapes today are maintained or altered by some combination of ecosystem and cultural processes.

Indeed, land managers, in deciding how to achieve environmental and social goals, may rely primarily on cultural processes or on ecological processes to maintain, restore, or enhance landscapes. The resulting landscapes would be very different (city versus forest) and would therefore provide different combinations of environmental and economic benefits, risks, and opportunities. Direct maintenance costs for landscapes maintained by ecological processes are, by definition, low. The direct costs of maintaining a landscape by cultural processes will be higher. In general, reliance on cultural processes will require ongoing effort (and expense) to prevent succession, decay, erosion, sediment deposition, and other ecological or physical processes from changing the landscape in undesirable ways. In addition, landscapes maintained primarily by cultural processes often change the rate of flow of sediments, nutrients, water, and other chemicals to adjacent ecosystems, perturbing them in unplanned and often undesirable ways. Thus, human-dominated landscapes often induce environmental externalities.

To understand why externalities arise, it is necessary to consider landscapes as hierarchical systems. Landscapes have usually been managed primarily at small scales; less attention has been paid to the effects of small-scale land use changes on larger-scale landscape and watershed processes, perhaps because the benefits to be derived are often public goods. The intercalation of cultural processes into the landscape hierarchy therefore is dominated by social, economic, and behavioral processes occurring at certain characteristic scales. At larger scales, we just do not pay much attention until some sort of a problem develops.

Within a dynamic hierarchy, however, changes in dynamics at specific scales are communicated up and down the hierarchy. Altering dynamic processes at one scale will trigger unplanned effects at other scales. One manifestation of this is that landscapes maintained predominately by cultural processes often export environmental problems to other landscapes with which they are linked (for example, the agricultural landscape of the Nanticoke exports nutrients to the estuary). These problems represent conditions imposed on adjacent ecosystems inconsistent with their previous ecologically driven dynamic regimes. Establishment of culturally maintained landscapes at one scale often has the effect of disrupting ecosystem maintenance processes at larger scales. At the smaller scale, culturally controlled landscapes may be of high quality (toward the top and right of the Ternary Diagram), yet at the next higher level in the landscape hierarchy, the landscape begins to change. At least in the short term, it is likely that some of these unplanned and often unanticipated landscape changes will be undesirable.

In the Chesapeake watershed, small-scale landscape transformation has been going on for centuries, with different results in different areas. In suburban or agricultural landscapes, as elsewhere, we have just begun to institute cultural processes aimed at maintaining the larger (watershed scale) landscape system. It should, therefore, not come as a surprise that the Chesapeake watershed shows signs of dynamic instability and functional change. In this case, many of the changes we see have been unpleasant, with the result that a concerted effort has developed to manage the Bay to maintain more of its environmental benefits, either by shoring up disrupted ecosystem processes or by implementing new cultural processes to partially replace them.

Management of Chesapeake Landscapes

Consider the management options managers of the Chesapeake Bay and its watershed face in hoping to restore environmental services once provided by the Chesapeake.

1. The Bay may be managed in a manner that relies on cultural processes to assist in the production of desired environmental outputs (for example, oyster aquaculture, rockfish hatcheries). We would call this approach environmental enhancement, since it would not be restoring the Bay system to previous dynamic conditions, but instead working to replace them with a new dynamic regime. Such a strategy is likely to become increasingly difficult and expensive as ecosystem services of the Bay watershed continue to decline.
2. Alternatively, the Bay could be managed to reinstate fundamental ecological relationships that previously

produced desired environmental benefits. This approach could be called environmental restoration because it focuses on restoring ecological processes that previously dominated the dynamics of the watershed. Environmental restoration, in this sense, may place more stringent limits on the scale or character of human activities within the Chesapeake watershed, inducing a variety of social costs. Different approaches or combinations will be most feasible in each landscape.

Environmental restoration is likely to play a significant role in agricultural landscapes such as the Nanticoke, where land use is relatively stable and the physical structure of the watershed is less profoundly disturbed. Impervious surfaces are not widespread in the Nanticoke watershed. The basic topographic structure necessary to reestablish ecological linkages among agricultural fields, drainage systems, and largely intact (but dewatered) riparian areas remains in place. Restoring wetlands and riparian areas would reestablish biological processing of nutrients and other chemicals. Such an effort would require that water be retained on the landscape for longer periods of time than is now the case. Increased retention would increase the exposure of both ground and surface waters to biological processing, and reestablish and maintain conditions in the soil conducive to denitrification. While restoring the hydrology of the Nanticoke watershed to something resembling its historical condition may be technically feasible, such a change would be expensive and probably unacceptable to local inhabitants. To be practical, restoration of landscape functions must be carried out in such a way that farmers' and other residents' need for drainage can continue to be met, while simultaneously retaining more water on the landscape, at least at some locations and at some times of year.

Much of the Nanticoke watershed once consisted of seasonal wetland systems, wet in the late winter and spring, but drier once evapotranspiration of growing trees and other plants removed water from the poorly drained soils. Thus, a major goal of many drainage systems was to remove water early in the growing season in order to ensure that agricultural lands could be planted. Ironically, some areas are now irrigated during the summer because they lack sufficient water to maintain plant growth. The ditches that remove water in the spring also reduce the availability of water the rest of the year.

Thus, while full restoration of natural hydrology is impractical for the Nanticoke, partial restoration of ecological functions may be possible using simple water control structures that can be opened a certain times of year, and closed at other times. Installation of such structures within selected ditches may allow removal of water from these lands in the spring (when the control structures are open), while permitting increased retention of waters in the floodplains during the remainder of the growing season, when agricultural as well as riparian and wetland areas could benefit.

In contrast to the rural landscapes of the Nanticoke watershed, few opportunities for large-scale landscape restoration are likely within urban and suburbanizing landscapes such as those in Anne Arundel County. Tree plantings and other typical urban and suburban restoration efforts build public awareness of environmental issues, but often have only marginal impacts on landscape-level processes. Larger-scale efforts at restoration are generally precluded because they would require the displacement of high-value land uses and removal of existing structures. Instead, most opportunities for managing landscape processes in suburban landscapes rely on enhancement of environmental functions with engineered structures like stormwater management devices.

Over the last two decades, stormwater management has been used increasingly to protect streams and surface waters from the consequences of local land use change (e.g. Schueler 1987). Infiltration, detention, and retention basins have become a fixture of the Maryland landscape, found wherever recent development has occurred. Stormwater management substitutes engineered structures designed to provide particular hydrologic and water quality services for physical and ecological processes disrupted by development. Installation of stormwater management structures represents a prime example of a landscape being managed to achieve environmental purposes in a way that increases reliance on cultural processes.

The potential value of stormwater management is great, but it comes with substantial costs. The engineered structures being used for stormwater management are costly and require ongoing maintenance to protect their environmental function. Maintenance shortcomings are common (Roberts and Lindsey 1990). Even when properly maintained, most stormwater management structures have a variety of unintended side effects (Schueler and Galli 1991). Infiltration devices, for example, not only infiltrate water but also inject dissolved pollutants into the ground water, where biological contact and treatment are low. Detention and retention basins increase surface water temperatures. In general, high maintenance requirements and a variety of environmental side effects are to be expected whenever cultural maintenance processes substitute for ecological ones.

Recent developments in stormwater management have tried to reduce or eliminate many of these problems (Schueler and Galli 1991). One answer has been to design stormwater management structures as shallow, vegetated wetlands (Moshiri 1993; Schueler 1992). These basins not only provide the water quality and quantity control required of stormwater management devices by state laws, but also provide an artificial context for ecological processes in a largely human-dominated landscape where such processes would otherwise be rare. Using artificial wetlands in this way reflects a growing understanding that environmental technologies can be most effective when built to exploit, rather than resist, ecological processes (Mitsch and Jorgensen 1989). On a larger scale, these artificial wetlands represent an effort to build a hybrid landscape that is neither natural nor artificial, but in which important natural processes are sustained in the context of a landscape that provided for human wants and needs. Such hybrid management systems, part nature, part culture, will, we suspect, become ever more common as our society learns to reconcile the self-sustaining character of ecosystems and the focused functionality of manufactured artifacts, and equally important, as we learn to recognize how human activity affects landscape-scale systems.

Conclusions

Residents of the Chesapeake Bay watershed have made, and continue to make, many land management decisions based on local benefits and short-term needs. Cumulatively, these decisions have provided food, housing, and other direct benefits for many, but have simultaneously altered landscape systems and initiated changes in landscape functions that reduced other benefits provided by the Bay, its watershed, and its tributaries.

We suggest that long-term restoration and protection of the Chesapeake Bay and its many values will require recognition of the hierarchical properties of landscapes. In the long run, successful management of the Chesapeake will require tools and approaches to environmental management able to assess landscape functions at various scales and to recognize cumulative effects of apparently isolated decisions. Efforts should be expanded to inform individuals how their actions impact the landscape, as well as to communicate what can be done to achieve individual land management goals while avoiding or mitigating negative impacts on landscape systems. Landscape-scale management should also provide feedback to individuals and local governments so that their actions can be better coordinated to reduce threats to landscape functions. In the long run, economic, legal, and other incentives and disincentives may have to be tailored to reduce local activities that create negative externalities at landscape scales, and to encourage those that support landscape processes.

Some landscape-centered management approaches already exist. For example, local governments and soil and water conservation districts which oversee land management within political boundaries have initiated various programs to promote land management practices that help to maintain landscape-level functions. Federal, state, and local regulations have also been adopted that coerce land managers to reduce certain impacts on landscape-scale systems and to provide mitigation for impacts that do occur. Such regulations, however, are facing strong opposition because they generally impose direct costs on individual land owners.

Interestingly, recent changes in efforts to manage the Chesapeake watershed also reflect the need for management that better reflects the hierarchical nature of landscape dynamics. The Chesapeake Bay Program, a regional cooperative effort to study and manage landscape-scale processes was established by the federal government and the main Chesapeake Bay states (Pennsylvania, Maryland, Virginia, as well as the District of Columbia) as a way to better study and manage landscape-scale problems that have led to loss of benefits from the Bay. Recently, leaders of the Chesapeake Bay Program initiated the "Tributary Strategies" which decentralize the watershed restoration effort and focus attention on the peculiarities of the different sub-watersheds of the Chesapeake. Coalitions of citizens, local governments, and soil and water district officials have been created to evaluate specific problems within the watersheds of the major Chesapeake

Bay tributaries, and to promote regional management at a landscape scale. A major component of this initiative is education of residents so that they may make land management decisions informed by how they impact the larger landscape system.

In the Chesapeake Bay watershed, re-establishing landscape processes able to support a healthy, more productive Bay will require recognition of and investment in landscape-scale processes. We will have to be far more sophisticated landscape and watershed managers than we are today if we are to support anticipated human populations of 15 million or more in the watershed by the year 2020 without causing further declines in the physical, chemical, and biological integrity of the Bay and its tributaries. In the future, management efforts that focus explicitly on enhancing or restoring watershed- or landscape-scale processes must become a central part of all efforts to protect the environmental benefits of the Chesapeake. Otherwise, the combination of the landscape-scale externalities associated with land use decisions and the limited ability of markets to efficiently allocate public goods will lead to continued deterioration of the Chesapeake watershed, unnecessarily impoverishing the region.

References

2020 Panel. 1988. *Population Growth and Development in the Chesapeake Bay Watershed to the Year 2020.* The Report of the Year 2020 Panel to the Chesapeake Executive Council.

Ailstock, S. M. and S. G. Horner. 1991. *Management Strategies for Urban Stormwater Improvement—District of Columbia Oxon Run, 13th to 22nd Street.* District of Columbia Housing and Environmental Regulation Administration. U.S.E.P.A. Environmental Center, Anne Arundel Community College, Arnold, MD.

Bohlen, C. C. 1992–93. Wetlands politics from a landscape perspective. Maryland *Journal of Contemporary Legal Issues* 4(1):1–12.

Bohlen C. C. and D. M. King. 1995. Location and wetland values: Some pitfalls of offsite wetland mitigation in the Chesapeake Watershed. In Steve Nelson and Paula Hill, (eds). *Towards a Sustainable Coastal Watershed: The Chesapeake Experiment.* Proceedings of a Conference. CRC Publication No. 149. Chesapeake Research Consortium, Edgewater, MD.

Bormann, F. H. and G. E. Likens. 1979. *Pattern and Process in a Forested Ecosystem.* Springer, New York, NY.

Brush, G. S. and F. W. Davis. 1984. Stratigraphic evidence of human disturbance in an estuary. *Quaternary Research* 22:91–108.

Chesapeake Bay Program. 1994. *Nutrient Reevaluation Report.* EPA Chesapeake Bay Office, Annapolis, MD.

Cooksey, R. 1993. *Forest Land Use Data Analysis.* Document Prepared for the Forestry Working Group of the Living Resources Subcommittee of the Chesapeake Bay Program.

Cooper, S. R. and G. S. Brush. 1991. Long-term history of the Chesapeake Bay anoxia. *Science* 254:992–996.

Environmental Protection Agency. 1991. *Proposed Guidance Specifying Management Measures for Sources of Nonpoint Pollution in Coastal Waters.* U.S. Environmental Protection Agency, Washington, DC.

Forman, R. T. and M. Godron. 1986. *Landscape Ecology.* John Wiley and Sons, New York, NY.

Grumbine, E. R. 1994. What is ecosystem management? *Conservation Biology* 8(1):27–38.

Hall, L. W. Jr., A. E. Pinkney, L. O. Horseman and S. E. Finger 1985. Mortality of striped bass larvae in relation to contaminants and water quality. In Hall, L. W. Jr., S. A. Fischer, W. D. Killen Jr., M. C. Scott, M. C. Ziegenfuss, and R. D. Anderson. 1994. Status assessment in acid sensitive and non-acid sensitive Maryland coastal plain streams using an integrated biological, physical and chemical approach. *Journal of Aquatic Ecosystem Health* 3:145–167.

Hall, L. W. Jr., S. A. Fischer, W. D. Killen, Jr., M. C. Zeigenfuss and R. D. Anderson. 1994. *1992 Doser Study in Maryland's Coastal Plain: Use of a Limestone Doser to Mitigate Stream Acidification.* Final Report to Maryland Department of Natural Resources Chesapeake Bay Research and Monitoring Division. Report Number CBRM-AD-94-2.

Hamilton, P. A. and R. J. Shedlock. 1992. *Are Fertilizers and Pesticides in The Ground Water? A Case Study of the Delmarva Peninsula, Delaware, Maryland, and Virginia.* U.S. Geological Survey Circular 1080. U.S. Geological Survey, Denver, CO.

Holling, C. S. 1992. Cross-scale morphometry, geometry and dynamics of ecosystems. *Ecological Monographs* 62:447–502.

Horton, T. and W. M. Eichbaum. 1990. *Turning the Tide: Saving the Chesapeake Bay.* The Chesapeake Bay Foundation. Island Press, Washington, DC.

Karr, J. R., L. A. Toth, D. R. Dudley. 1985. Fish communities of midwestern rivers: A history of degradation. *BioScience* 35(2):90–95.

Klein, R. D. 1979. Urbanization and Stream Quality Impairment. American Water Resources Association. *Water Resources Bulletin* 15(3): 948–963.

Lackey, R. T . 1994. *The Seven Pillars of Ecosystem Management.* Report given at "Ecosystem Health and Medicine: Integrating Science, Policy, and Management." Ottawa, Ontario, Canada, June 19–23.

Linker, L. C., G. E. Stigall. C. H. Chang, and A. S. Donigion. 1995. *Chesapeake Bay Watershed Model.* Accepted for publication in Water Environment and Technology.

Lowrance, R., L. S. Altier, J. D. Newbold, Ronald R. Schnabel, Peter M. Groffman, Judith M. Denver, David L. Correll, J. Wendell Gilliam, James L. Robinson, Russell B. Brinsfield, Kenneth W. Staver, William C. Lucas, and Albert H. Todd. 1995. *Water Quality Functions of Riparian Forest Buffer Systems in the Chesapeake Bay Watershed.* Technology Transfer Report of the Chesapeake Bay Program. Nutrient Subcommittee.

Maryland Department of the Environment (MDE). 1992. *Maryland's Ozone Problem.* Handouts from a presentation to the Baltimore Metropolitan Transportation Steering Committee. Maryland Department of the Environment, Baltimore, MD.

Maryland Department of the Environment (MDE). 1993. *Maryland Tributary Strategies: Restoring the Chesapeake, Overview.* February 1993. Maryland Department of the Environment, Baltimore, MD.

Maryland Department of the Environment (MDE). 1994. *Chesapeake Bay Tributary Monitoring Reports: Nanticoke River: Tidal Fresh.* August 1994 draft. Maryland Department of the Environment, Baltimore, MD.

Maryland Office of Planning. 1991. *Maryland's Land. 1973–1990: A Changing Resource.* Maryland Office of Planning, Annapolis, MD.

Maryland Office of Planning. 1993. *Areas Experiencing Growth and Decline in the State of Maryland 1970–1990.* Map Prepared by Maryland Office of Planning Data Services Division. February 1993.

McKibben, B. 1989. *The End of Nature.* Anchor Books, Doubleday, New York, NY.

Mitsch, W. J. 1992. Landscape design and the role of created, restored, and natural riparian wetlands in controlling nonpoint pollution. *Ecological Engineering* 1(1):27–48.

Mitsch, W. J. and S. E. Jørgensen. 1989. *Ecological Engineering: An Introduction to Ecotechnology.* John Wiley and Sons, NY.

Moshiri, G. A. 1993. *Constructed Wetlands for Water Quality Improvement.* Lewis Publishers, Boca Raton, FL.

Nature Conservancy. 1994. *Nanticoke-Blackwater River Bioreserve Strategic Plan.* Draft 6-10-94. Maryland Nature Conservancy, Chevy Chase, MD.

Neumiller, S. K., L. C. Linker, J. E. Hannawald, A. S. Donigian, Jr., and B. R. Bicknell. 1994. *Phase III Watershed Model Application To Calculate Bay Nutrient Loadings: Final Findings and Recommendations.* Appendix E: Land Use and Selected Parameter Values. Chesapeake Bay Program, U.S. Environmental Protection Agency, Annapolis, MD.

O'Neill, R. V., D. L. DeAngelis, J. B. Waide, and T. F. H. Allen. 1986. *A Hierarchical Concept of Ecosystems.* Monographs in Population Biology 23. Princeton University Press, Princeton, NJ.

Odum, E. P. 1969. The strategy of ecosystem development. *Science* 164:262–270.

Olsenholler, S. M., 1991. *Annual Loading Estimates of Urban Toxic Pollutants in the Chesapeake Bay Basin.* Department of Environmental Programs, Metropolitan Washington Council of Governments, Washington, DC.

Peterjohn, W. T. and D. L. Correll. 1984. Nutrient dynamics in an agricultural watershed: Observations on the role of a riparian forest. *Ecology* 65(5):1466–1475.

Preston, E. M. and B. L. Bedford. 1988. Evaluating cumulative effects of wetland functions: A conceptual overview and generic framework. *Environmental Management* 12:565–583.

Richardson, C. J. 1994. Ecological functions and human values in wetlands: A framework for assessing forestry impacts. *Wetlands* 14(1):1–9.

Roberts, L. and G. Lindsey. 1990. *Maintenance Requirements for Stormwater Facilities in Baltimore County.* Sediment and Stormwater Administration, Maryland Department of the Environment, Baltimore, MD.

Schlosser, I. J. 1991. Stream fish ecology: A landscape perspective. *BioScience* 41(10):704–712.

Schueler, T. R. 1987. *Controlling Urban Runoff: A Practical Manual for Planning and Designing Urban BMPs.* Metropolitan Washington Council of Governments, Washington, DC.

——— . 1991. Mitigating the adverse impacts of urbanization on streams. pp. 21–31 in Anacostia Restoration Team, *Watershed Restoration Source Book—Collected Papers.* Presented at the conference "Restoring Our Home River: Water Quality and Habitat in the Anacostia." Nov. 6–7, 1991. Metropolitan Washington Council of Governments, Washington, DC.

———. 1992. *Design of Stormwater Wetland Systems: Guidelines for Creating Diverse and Effective Stormwater Wetland Systems in the Mid-Atlantic Region.* Metropolitan Washington Council of Governments, Washington, DC.

Schueler, T. R. and J. Galli. 1991. The environmental impact of stormwater ponds. pp. 161–180 in Anacostia Restoration Team, *Watershed Restoration Source Book—Collected Papers.* Presented at the conference "Restoring Our Home River: Water Quality and Habitat in the Anacostia." Nov. 6–7, 1991. Metropolitan Washington Council of Governments, Washington, DC.

Tiner, R. W. Jr. 1987. *Mid-Atlantic Wetlands—A Disappearing Natural Treasure.* U.S. Fish and Wildlife Service, Fish and Wildlife Enhancement, National Wetlands Inventory Project, Newton Corner, MA.

U.S. Department of Commerce. 1954, 1959, 1964, 1969, 1974, 1982, 1987, 1992. *Census of Agriculture: State and County Data for Delaware.* Bureau of the Census, Washington, DC.

U.S. Department of Commerce. 1954, 1959, 1964, 1969, 1974, 1982, 1987, 1992. *Census of Agriculture: State and County Data for Maryland.* Bureau of the Census, Washington, DC.

U.S. Department of the Interior. 1994. *The Impact of Federal Programs on Wetlands.* Vol. II. A report to Congress by the Secretary of the Interior. Washington, DC.

Welsch, D. J. 1991. *Riparian Forest Buffers: Function and Design for Protection and Enhancement of Water Resources.* U.S. Dept. of Agriculture, Forest Service North Eastern Area, State And Private Forestry, Forest Resource Management. Radnor, PA. Report Number NA-PR-07-91.

CHAPTER SIX

History and Impact of Human Activities on Chesapeake Bay

Grace S. Brush

Introduction

This paper investigates the role humans have played in changing the connections between the Chesapeake Bay estuary and watershed; during the process, both systems have been radically altered. The paleoecological record of indicator organisms and materials preserved in estuarine sediments has been used to reconstruct the history of the estuary and the land it drains, thus providing a predictive yardstick for ecosystem restoration.

The Chesapeake Bay originated some 10 to 12,000 years ago when the latest continental ice sheet, which extended to the approximate vicinity of Scranton, Pennsylvania, began to recede northward. The resulting rise in sea level flooded the coastal region, including the Susquehanna River Valley, forming an estuary named the Chesapeake Bay. Within that 10,000-year span, climate changed from cold to warm and varied also between wet and dry. When Europeans emigrated to this region a few centuries ago, they found a totally forested landscape inhabited by sparse Indian populations who were mainly hunters and fishermen. Within less than a century, almost half of the land surrounding the Chesapeake Bay was under cultivation for various agricultural crops.

The transformation from a forested to a non-forested landscape was accompanied by the conversion of a diverse estuary, dominated by benthic processes, into one dominated by plankton. In shallow water systems, where light can penetrate the water column, the benthos can be highly productive, and so it was with the Chesapeake Bay during its lifetime, despite drastic shifts in climate. The remarkable thing is that the reversal which began a little over a century ago was not recognized until recently, when there were precipitous declines in the commercial fishery. But the signs of impending disaster were clearly recorded in the sediments. The slow and steady pattern of decline leading to dramatic shifts in species composition and abundance reflects the nature of biological systems, which can adjust to a deteriorating habitat for a while, but at some level of stress reach a critical threshold, when sensitive populations clash with many species becoming locally extinct.

The ecosystem we perceive is not the result of instantaneous events, but has a long history related to geology and climate. The response of ecosystems to natural conditions determines their sustainability with regard to the amount and kind of bioresources they can provide. This capacity changes over time, because temperature and precipitation, which are primary factors in the manufacture of biomass, are climatically controlled and highly variable even at a single location. Availability of nutrients, also important in biomass production, is related to the physical and chemical conditions of the soils which change as weathering, soil formation, and soil erosion proceed. And the estuarine substrate also controls nutrient recycling, and hence availability of nutrients within the estuary.

Another important aspect of ecosystems is the interrelationships between them. None are closed systems. What happens in one ecosystem is likely to affect other ecosystems within the same watershed. Changes on the land can profoundly affect the aquatic system which drains the land. Atmospheric pathways connect spatially separated ecosystems, even over long distances.

The stratigraphic record shows a fundamental difference between the response of estuarine populations to change that is climatically induced, and that resulting from anthropogenic land use. The former is episodic and the latter continuous. Organisms living in a dynamic environment such as was characterized by the pre-European Chesapeake adapt to periodic catastrophes by becoming dormant during unfavorable times, and evolving life cycles adapted to tidal, seasonal, and longer-term climate changes that shape and characterize their habitats. Land use as practiced in the Chesapeake region, as well as in other parts of North America, was so intense that it permanently changed aquatic habitat characteristics, particularly the benthic habitats. The organisms simply could not handle the physiological stress accompanying what was actually a massive loss of habitat.

For the estuarine system to be sustainable, the benthic habitats must be restored. Because land use and estuarine sustainability are so closely related, this can only be done by making the altered landscape more closely resemble the natural forested condition.

In order, therefore, to design management protocols that in the face of human exploitation will allow ecosystems to operate at their maximum capacity, recognizing that their maximum capacity varies as processes related to climate and geomorphology change, it is essential to understand their dynamics and long-term history.

It is difficult to extract long-term history from written records, because historical records of environmental conditions and ecosystem production are not available for any particular area over a sufficiently long time period to capture, and place within the context of time, the major changes affecting the system. But the surrogate history contained in sediments deposited in aquatic systems provides detailed quantitative information from which generalizations can be made with regard to sustainability. Furthermore, these stratigraphic records extend over time periods that precede European settlement, and thus allow a comparison of the system when changes were primarily the result of climatic events with the period of time when anthropogenic activity became the dominant influence.

This surrogate stratigraphic record has been used to trace the history of the Chesapeake Bay and to reconstruct the influence of land use on the estuarine resource.

General Description of the Area of Study

The Chesapeake Bay is located in the mid-Atlantic region of the USA between 37° and 39°40' North latitudes (Figure 6–1). It is about 290 kilometers long and drains a total area of approxi-

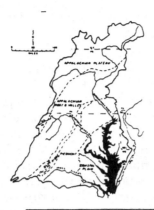

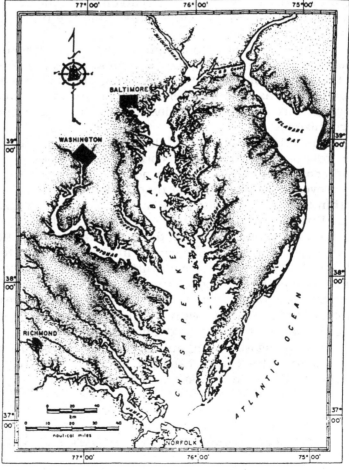

Figure 6–1. Map of the Chesapeake Bay with inset showing the watershed.

mately 166,000 square kilometers. It is the largest estuary in North America and the fourth largest in the world. The human population occupying the Chesapeake watershed has grown from 2 million when Europeans emigrated to North America to 14 million today. Prior to European settlement, the entire area was forested, except for some tidal wetlands and barren areas on serpentine rock. Native populations were small in number and relied mainly on hunting and fishing for their livelihood. The present resources and condition of the Chesapeake Bay are described by Boynton (this volume).

The Stratigraphic Record

The surrogate historical record of environmental history is derived from the stratigraphy of the sediments. The hydrodynamics of the Chesapeake Bay indicate that it acts as a large settling basin where small particles, once they enter the estuary, are not transported far before they are deposited (Figure 6–2) (Brush and Brush 1994). As silt and clay particles are deposited, they bury organisms, parts of organisms, chemicals, etc., that occur in or are transported into the estuary by wind and water. If the burial is reasonably rapid, most biological and many chemical components are preserved in the sediment, providing a history of environmental conditions over the period of time during which the sediment was deposited. This alternate historical record is retrieved by collecting sediment cores and analyzing them for those organisms and chemicals that are indicators of water quality and environmental conditions.

Sediment cores collected throughout the Chesapeake Bay and its tributaries are dated using carbon-14 and pollen analyses. The analyses provide sedimentation rates for samples within the core, thus allowing sediment depth to be converted to time in years, and providing chronological records of events and change analogous to historical documents. The samples from the cores are archived and thus available for future analyses. Sedimentation rates also allow the conversion of concentrations of the paleoecological components to influxes or the amount deposited per year. The highly variable sedimentation rates within the estuary and tributaries make it necessary to compare influxes of materials rather than concentrations. For example, 100 diatoms in 1 cm^3 of sediment could mean that 10 diatoms were

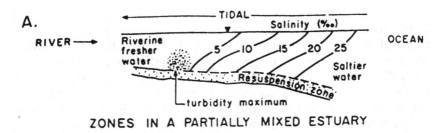

ZONES IN A PARTIALLY MIXED ESTUARY

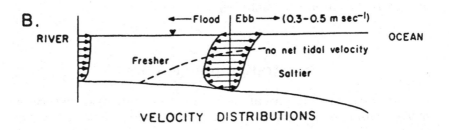

VELOCITY DISTRIBUTIONS

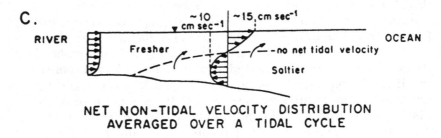

**NET NON-TIDAL VELOCITY DISTRIBUTION
AVERAGED OVER A TIDAL CYCLE**

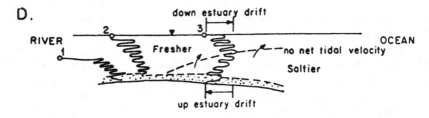

Figure 6–2. Diagram of hydrodynamics of the Chesapeake Bay (from Brush and Brush 1994). a. the salinity zones in the Bay showing the turbidity area and the zone of resuspension; b. the velocity distributions; c. the net non-tidal velocity distribution averaged over a tidal cycle; d. the likely fate of particles like silt, clay and pollen which enter the estuary at locations 1, 2 and 3.

deposited in one year if the cm³ of sediment were deposited in ten years, whereas if the cm³ were deposited in one year, the actual number of individuals deposited in one year would be 100.

Table 6–1 lists some components used in deciphering the history of the Chesapeake Bay, but there are others such as beetles, copepods, insects, fish scales and plant pigments which have not been utilized thus far in our studies.

Table 6–1   Fossil indicators and their use in interpretation of environmental history

Fossil component	Indicator of	Interpretation
Pollen and seeds of terrestrial habitats	Availability of water in soil	Climate (precipitation)
Pollen and seeds of submerged and emergent vegetation	Occurrence of submerged aquatic vegetation and emergent vegetation	Occurrence of nursery grounds for shellfish and finfish; indicator of water levels; indicators of filling in from sedimentation or also of changes in sea level
Pollen and seeds of marsh plants	Indicator of type of marshsalt or fresh, high or low	Climate (sea level); precipitation; deforestation (erosion)
Rhizomes	Indicators of marsh plants	Marsh environment
Benthic diatoms	Salinity; light availability; available oxygen	Land clearance (turbidity); eutrophication (fertilizers and sewage); anoxia
Planktonic diatoms	Available light	Turbidity; eutrophication
Total organic carbon, nitrogen, sulfur, phosphorus	Benthic oxygen demand; eutrophication	Anoxia; nutrient loading and reductions
Charcoal	Fire	Climate (severe drought); anthropogenic fire
Trace metals	Fire; toxics	Climate (fire); toxic pollution
Mass sediment	Sediment accumulation; turbidity	Deforestation from anthropogenic activity and fire

History of Climate Change

The longest continuous record of postglacial vegetation found thus far in the mid-Atlantic region is contained in sediments deposited in the floodplain of Indian Creek, a tributary of the Anacostia River close to Washington, D.C. (Yuan 1995) (Figure 6–3). The sequence which began 12,000 years ago shows a coniferous forest consisting predominantly of fir, spruce, and pine lasting for about 2,000 years. Other species in this assemblage include alder, ash, birch, hornbeam, and hazelnut. There was very little oak and no hickory. Grasses and sedges dominated the non-arboreal flora. Climate was cold and wet. This boreal type of forest was succeeded by hemlock, which dominated the landscape for 5,000 years. Black gum appeared at this time and gradually increased. The predominance of pine and hemlock with some deciduous trees such as black gum, alder, and birch, but with few herbaceous plants, indicates a closed canopy mixed coniferous-deciduous forest. After 5,000 years, large amounts of charcoal in the sediment, accompanied by oak and hickory, indicate a warmer and drier climate characterized by frequent fires. The oak-hickory forest spanned 3,500 years prior to European settlement. During this interval, while oak and hickory were increasing, pine decreased. Sweet gum appeared on the landscape. Herbaceous plants increased and members of the blueberry family emerged as a major component of the vegetation. Ragweed constitutes a large portion of the pollen flora in sediments deposited in the last 350 years. Ragweed is a native annual plant which colonizes exposed mineral soil. An abundance of ragweed pollen in the sediment is an indicator of secondary succession following cultivation of the land for agriculture.

The history of the past few thousand years has been reconstructed from roughly 100 sediment cores collected throughout the Chesapeake Bay and tributaries. These cores span European occupation of the area, the Little Ice Age (LIA) which extended from the thirteenth into the nineteenth century, and the Medieval Warm Period (MWP) from about 800 or 900 to 1200 A.D. These two climatic intervals which preceded European settlement in North America are well documented by historical, stratigraphic, and tree ring data in both Europe and the United States.

Temperatures worldwide were anomalously high for approximately two to five centuries from about 800/900 to 1200/1300

Pollen percentages from Dan's Bog, Maryland

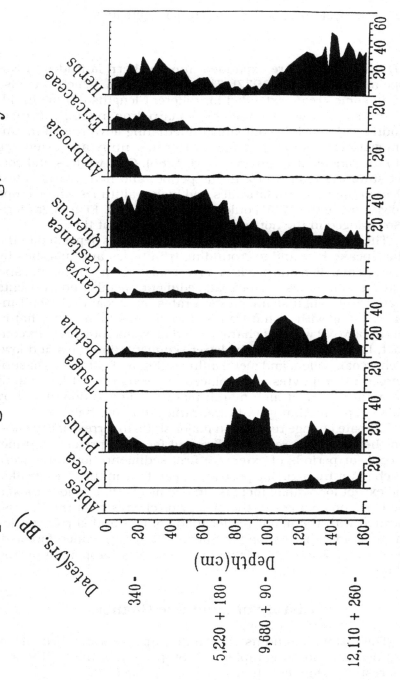

Figure 6–3. A pollen profile from Dan's Bog located on the floodplain of Indian Creek, a tributary of the Anacostia River close to Washington, D.C., showing the history of vegetation in this area since the time of glacial retreat some 12–15,000 years ago. (Yuan 1995)

A.D., and agriculture expanded into northern latitudes in Europe and Asia (Lamb 1977). This warm period, which was also dry in some areas, extended for different lengths of time in different places. It was followed by a sharp drop in temperature of around 2°C which accompanied fluctuating advances of mountain glaciers, beginning in the thirteenth century and extending into the nineteenth century (Grove 1988). Cold winters and cool wet summers characterized this period, and had their greatest effect in the northern latitudes and higher altitudes where crops grown during the MWP could no longer survive. Fish migrations also changed in the northern waters during that time.

The MWP and LIA are recorded in sediments deposited in the Chesapeake Bay and surrounding tributaries and marshes by changes in pollen and seeds of terrestrial and aquatic plants, and changes in influxes of charcoal, sediment, metals and nutrients (Figure 6–4). High sedimentation rates, charcoal and metal influxes, along with a shift from wet to dry plants indicate that in this area MWP was dry characterized by numerous fires. In contrast, LA is characterized by lower sedimentation rates and lower charcoal, pollen, and metal influxes (Figure 6–4). Seeds of submerged macrophytes are replaced by seeds first of low marsh plants and later of high marsh plants. Pollen is dominated by pollen of plants that grow in wet rather than dry habitats.

Climate change resulted in major shifts in terrestrial species, but the landscape remained forested for all of prehistoric time. Except for periods of extensive fire, sedimentation rates were much lower than during post-European time in the Chesapeake, and except for a small increase in chemicals that denote anoxia, there are no changes in the biology or chemistry of the Bay associated with the major reallocations of tree and shrub species on the landscape. As long as the landscape remained forested, the estuary was little affected by what was happening on the land.

History of Land Use Change

The history of land use in the Chesapeake watershed shows that deforestation was rapid, with 80 percent of the land totally deforested within less than 150 years (Table 6–2).

The amount of land cleared at any time was not proportional to the size of the population. Population growth and land use

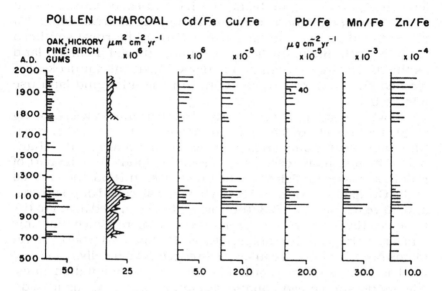

Figure 6–4. A stratigraphic profile of pollen, charcoal and metals extracted from Red Fin Creek on the Nanticoke River. (From Brush 1986) The profile shows the changes in the ratios of oak, hickory, and pine (dry species) pollen to birch, sweetgum and blackgum (wet species) during MWP about 1,000 years ago, as well as changes in the metal influxes.

Table 6–2   History of land use in the Chesapeake region

pre-European (pre seventeenth century)	entire area forested except for tidal wetlands and serpentine barrens
late seventeenth to mid-eighteenth centuries	20 to 30 percent of land cleared, much of it for tobacco farming
late eighteenth to mid-nineteenth century	40 percent of land cleared for small farms engaged in grain farming as well as tobacco
late nineteenth into the twentieth century	60 to 80 percent of land cleared for large farming operations; deep plough zone; fertilizers-guano in the 1870s followed by chemical fertilizers
early twentieth century	wetlands drained for agriculture
early to mid-twentieth century	60 percent of land under cultivation; decrease due to farm abandonment

in counties drained by the Patuxent River show a substantial difference (Figure 6–5a). In the northern counties, urbanization led to increased populations concentrated in small areas without a proportional increase in land use, whereas the rural southern counties with much smaller populations utilized as much land as the much larger urban populations. The rural Nanticoke watershed shows similar trends both in population and land use (Figure 6–5b).

It was recognized very early that tributaries were being clogged from soil-eroded sediment as deforestation and agriculture intensified. Commercial ships could not navigate the channels, which in many cases were becoming mudflats. The port of Baltimore was established on the Bush River in 1663, then moved to the Gunpowder River in 1693 where it was called Joppatowne, and 50 years later in 1723, the port had to be moved again to the Patapsco River, where it was renamed Baltimore. Since that time it has remained on the Patapsco River (Baltimore Harbor) and remains operable only because it is dredged periodically.

The stratigraphic record of changes in sedimentation rates, chemistry, submerged aquatic vegetation, and diatoms in sediment deposited over the past few centuries illustrates the enormity of the effect of land use on the Chesapeake Bay.

The sediment cores show that the average post-European sedimentation rate in the shallow main stem of the Chesapeake Bay is 0.17 cm/yr^{-1}, about three times greater than the average pre-European rate of 0.06 cm/yr. Average sedimentation rates for all tributaries studied are two times higher after European settlement (0.30 cm/yr^{-1}) than before settlement (0.14 cm/yr^{-1}). And the average upstream and midstream rates are 0.32 to 0.35 cm/yr^{-1}, twice as high as the average rate of 0.17 cm/yr^{-1} in the lower stretches, indicating that the major effect of land clearance was in the upper and middle sections of the tributaries (Brush 1984a). Sedimentation rates in Furnace Bay, a small embayment at the head of the Chesapeake Bay (Figure 6–6), reflect regional deforestation and agriculture, local charcoal and quarrying activities, and major storms (Brush 1989).

The effect of nutrient enrichment from fertilizers is seen in gradual increases in total organic carbon above what would be expected from natural degradation of carbon (Figure 6–7) (Cooper and Brush 1991). Cores analyzed for diatoms show a replacement of pennate diatoms with centric diatoms (Cooper 1993). The majority of pennate species in the Chesapeake Bay are ben-

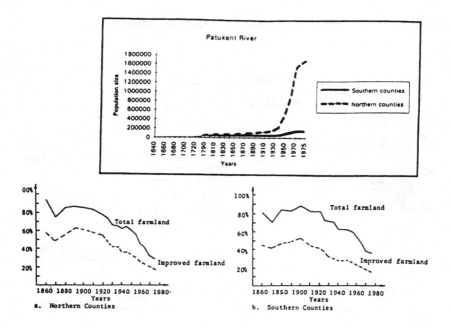

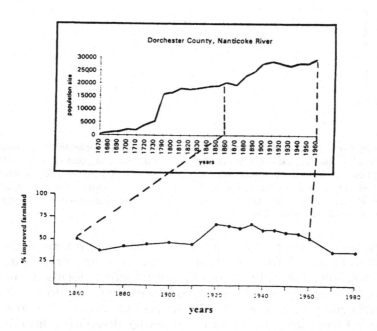

Figure 6–5. Population and land use trends in counties adjacent to a. the Patuxent River and b. the Nanticoke River. The northern counties of the Patuxent have become urbanized, while the southern counties of the Patuxent and all of the Nanticoke are rural.

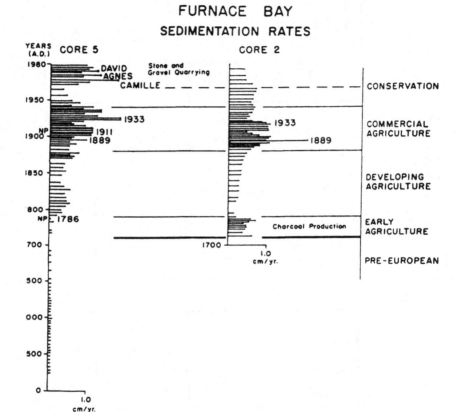

Figure 6–6. A profile of sedimentation rates from two cores separated by 0.5 kilometers in Furnace Bay located in the upper Chesapeake Bay. The patterns show that local land use such as quarrying and charcoal production affected only local areas. Similarly not all storms are reflected in all sediment cores, because storms have different tracks. The effects of agriculture are shown similarly in both cores because it is a regional phenomenon. (From Brush 1986)

thic; many are epiphytic, living on submerged macrophytes and macroalgae. Centric diatoms in the Bay are predominantly planktonic. The shift from pennate to centric forms during the past few centuries reflects in large part increases in the centric *Cyclotella spp.* (Cooper 1993). Chlorophyll influxes in cores from Back River mirror the volumes of sewage discharged into the river since 1911 (Figure 6–8), (Brush 1984b). In other tributaries where waste water is being discharged, *Cyclotella spp.* increased by an order of magnitude.

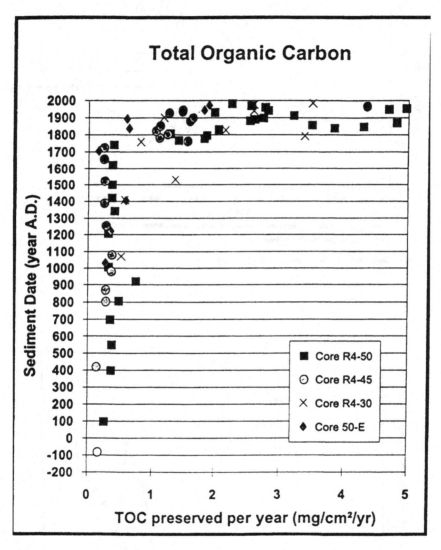

Figure 6–7. The stratigraphic profile of the amount of total organic carbon preserved per year in sediments in the main stem of the Bay (from Cooper 1993). TOC increased dramatically after European settlement in all 4 cores.

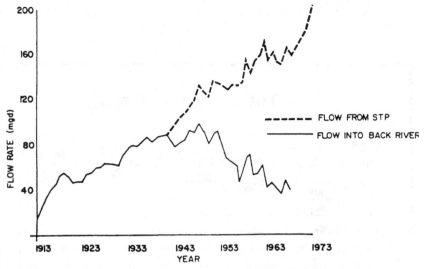

In 1943, part of the flow from the STP was diverted to Bethlehem Steel plant thus reducing flow into Back River.

Data taken from Simmons, D. CRC Report

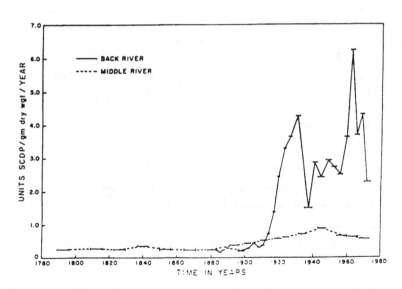

Figure 6–8. a. The pattern of sewage discharge into Back River from the Back River Sewage Treatment Plant and b. the profile of sedimentary chlorophyll preserved in the sediment from 1780 to the present. The profile of sedimentary chlorophyll from an adjacent tributary, Middle River, is shown for comparison. (Redrawn from Brush 1994)

A combination of high sedimentation rates, increasing eutrophication measured by total organic carbon and nitrogen, and
the production of hydrogen sulphide measured by the amount of
sulphur and iron in sediments deposited after European settlement, suggests that the present condition of anoxia, although
enhanced by fresh water inflow and intensified stratification, has
become more persistent with the introduction of increased nutrients into the water (Cooper and Brush 1991). The effect of sedimentation, eutrophication, and anoxia on benthic populations
is documented in the sediment cores by analyzing seeds of submerged macrophytes, the habitat for many estuarine species.

Seed influxes show highly variable fluctuations in submerged macrophyte populations both before and after European
settlement (Brush and Hilgartner, unpublished data). However,
increased numbers of seeds of all species are found in sediments
deposited immediately following colonization, a response to initial enrichment. As the waters became more turbid and light
more limiting with increased sediment and planktonic algal
growth, the submerged macrophytes were seriously stressed.
The severe declines and local extinctions, documented by aerial
photography and field surveys since the 1970s (Orth and Moore
1984), are registered in the stratigraphy by a reduction or absence of seeds of submerged macrophytes in sediments deposited since 1970.

The presence of a brackish species of submerged macrophyte, *Ruppia maritima*, in freshwater stretches at the heads of
the Chester and Severn Rivers up until the time of European settlement is strong evidence that those areas were more saline
than at present. The present distribution of *R. maritima* is restricted to downstream stretches of the tributaries, suggesting
that massive deforestation of the Chesapeake watershed resulted in a larger amount of fresh water entering the Bay through
river discharge and hence a less saline estuary.

The sequence of events is summarized as follows. Water in
the upper parts of the tributaries and shallow embayments became more turbid with increased sedimentation and eutrophication. As a result, there was insufficient light for the aquatic
grasses, which also were habitat for many of the epiphytic diatoms. The majority of the grasses died out. Benthic diatoms
were replaced by planktonic species, which could live in the upper water column where there was still sufficient light (Figure
6-9). Diatom abundance increased dramatically but diversity
was greatly reduced, with the majority of diatoms belonging to

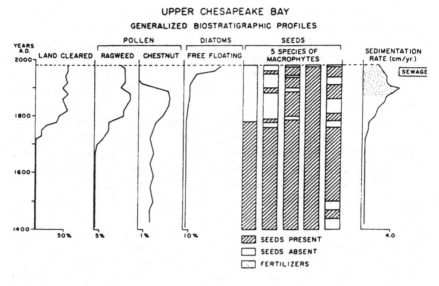

Figure 6–9. A summarized stratigraphic profile of changes in components of the terrestrial and estuarine ecosystems since 1400 A.D. Ragweed increased as soil was exposed for agriculture, and chestnut died out due to disease. (Redrawn from Brush 1986)

one planktonic species (Cooper 1993, 1995). Benthic fish populations, abundant at the time of colonization, were eventually depleted (Miller 1986). More recently, the shad, striped bass and oyster populations have also been greatly diminished.

Managing the Chesapeake for Sustainability

The stratigraphic history of the Chesapeake shows that shifts in the terrestrial flora in response to major climate changes were not reflected by corresponding changes in the estuarine diatoms and submerged macrophytes. There was a high degree of variability in the estuarine populations, but there were no changes in diversity and no large-scale extinctions. Decreased diversity, the shift from benthic to planktonic diatoms, and the extinction of much of the benthic submerged macrophyte community throughout the Bay all occurred after European settlement.

A comparison of sedimentation rates during MWP and the post-European period shows large increases during both periods, but there is much higher variability during the MWP (Figure

6–10). Soil erosion and sedimentation increased as fire frequency increased; in some cases the sedimentation rates could be as high as post-European time, but the pattern is episodic. Deforestation was not episodic but occurred with ever-increasing intensity as more land was cleared for agriculture. This is reflected by continuously increasing sedimentation rates.

Sustainability of the estuarine system will be restored only if the benthic habitats become productive again. In order for this

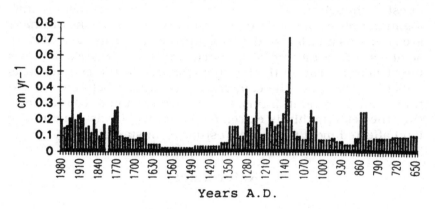

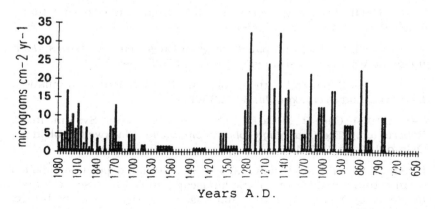

Figure 6–10. Sediment and charcoal profiles from Red Fin Creek, the Nanticoke River, showing large amounts of charcoal and sediment deposited in the river during the MWP. Both are substantially reduced during the subsequent LIA, and increase again after European settlement.

to happen, land use must mimic natural disturbance, where sedimentation accompanying land use is episodic. This can be accomplished by buffering the estuary as the pre-historic forests buffered it, which requires that the landscape itself be restored to something of its natural forested state. Estuarine degradation has proceeded in lockstep with the conversion of the forested watershed to a non-forested landscape, which was accomplished and is maintained by massive disturbance. Wherever farms are abandoned, and no other development has started, the land immediately reverts through a series of successional changes to forest. Although there has been some farm abandonment resulting in afforestation, still 40 percent of the Chesapeake drainage area remains non-forested. Managing the landscape so that it resembles a fine mosaic of forested and non-forested habitats might help to restore the benthic resource in the estuary. It is difficult to decipher the complex interrelations between the estuary and the land it drains, but the stratigraphic record shows that the relationship is extremely close; what was done to the land affected most profoundly estuarine sustainability.

References

Brush, G. S. 1984a. Patterns of recent sediment accumulation in Chesapeake Bay (Virginia-Maryland. USA) tributaries. *Chemical Geology* 44:227–242.

———. 1984b. Stratigraphic evidence of eutrophication in an estuary. *Water Resources Research* 20:531–541.

———. 1986. Geology and paleoecology of Chesapeake estuaries. *Journal of the Washington Academy of Sciences* 76(3):146–160.

———. 1989. Rates and patterns of estuarine sediment accumulation. *Limnology and Oceanography* 34:1235–1246.

———. 1994. Case 2. The Chesapeake Bay Estuarine System. *In:* N. Roberts (ed.), *The Changing Global Environment.* Blackwell Publishers, Oxford. UK, pp. 397–416.

Brush, G. S. and L. M. Brush. 1994. Transport and deposition of pollen in an estuary: a signature of the landscape. In A. Traverse (ed.), *Sedimentation of Organic Particles.* Cambridge University Press, Cambridge, UK., pp. 33–46.

Cooper, S. R. 1993. *The History of Diatom Community Structure, Eutrophication and Anoxia in the Chesapeake Bay as Documented in the*

Stratigraphic Record. Ph.D. dissertation, Johns Hopkins University, 301 pp.

———. 1995. Chesapeake Bay watershed historical land use: Impact on water quality and diatom communities. *Ecological Applications* 5(3):703–723.

Cooper, S. R. and G. S. Brush. 1991. Long-term history of Chesapeake Bay anoxia. *Science* 254:992–996.

Grove, J. M. 1988. *The Little Ice Age.* Metheun, New York, NY.

Lamb, H. H. 1977. *Climate; Past, Present, Future. Vol. II, Climate History and the Future.* Metheun, New York, NY.

Miller, Henry. 1986. Transforming a "splendid and delightsome land": Colonists and ecological change in the 17th and 18th century Chesapeake. *Journal of the Washington Academy of Sciences* 76(3):173–187.

Orth, R. J. and K. A. Moore. 1984. Chesapeake Bay: An unprecedented decline in submerged aquatic vegetation. *Science* 222:51–53.

Yuan S. 1995. *Postglacial History of Vegetation and River Channel Geomorphology in a Coastal Plain Floodplain.* Ph.D. dissertation, Johns Hopkins University, 164 pp.

CHAPTER SEVEN

Human Behavior and Ecosystem Valuation: An Application to the Patuxent Watershed of The Chesapeake Bay

Jacqueline Geoghegan

Nancy Bockstael

Introduction

The economic valuation of ecosystems has attracted the attention of both scientists and policy makers in recent years. However, most of the research in the past on this topic has been isolated within the individual paradigms of ecology and economics, with little interaction between the two. Unfortunately, this autonomous style of scholarship has led to fragmented and incomplete models. The validity of these models for policy implications can then be questioned.

Ecosystems are highly complex systems, with very few of the relationships between functions (described by the operational methodology) and form (which are the important characteristics of a system), as well as the impact humans have on function and form. There can be great uncertainty in the quantification of ecosystem functions such as life-support or pollution assimilative capabilities. However, even given these relationships, the relative importance of the impacts is unknown. At the same time, the knowledge of the consequence of ecosystems functions on human well being is also lacking. Most of the analysis has been limited to a small number of species and resources of the ecosystem that have obvious and immediate economic or human health impacts.

Ecologists have long criticized economists for their narrow and anthropomorphic view in the role, function, and importance of ecological systems by focusing on specific and immediate on-site impacts of human behavior on ecosystems. Economists have, in general, incorporated specific ecological functions into their models, ignoring the ecological concept that all things are interdependent. Ecologists have incorporated human behavior in their models as exogenous shocks to the ecological system, thereby over-simplifying the complex interaction between humans and ecosystems. Economists have also traditionally found fault with ecologists for their unwillingness to evaluate the relative contribution of different ecosystems services as well as for disregarding human preferences in considering value. Neither approach accounts for the interaction between the processes and populations each seeks to model.

Recognizing the drawbacks of isolated research, the U.S. Environmental Protection Agency (EPA) convened the "Ecosystem Valuation Forum" in order to bring ecologists and economists together to seek solutions and to combine efforts to overcome these shortcomings. The members of the forum stressed the importance of providing improved information to decision makers involved with ecosystem issues.

A research agenda to advance the sophistication of ecosystem valuation models was one of the results of this forum. Two major research areas were defined: (1) increasing the understanding of ecosystem function and how human behavior impacts those functions; and (2) improving the identification and methods of determining which ecosystem services are of value to humans (Bingham et al. 1995).

Many ecosystem functions do not provide a direct service to people, but influence attributes of the ecosystem that are important to humans. More information on these linkages is needed. In addition, a single ecosystem function or component can play more than one role in providing values to humans. An example of this is the role oysters play in the environment. Oysters provide obvious value to those who enjoy consuming them, but they also provide a service to the ecosystem and population at large by filtering and cleaning the water in which they live.

As we learn more about ecosystems, possible "new" valuable goods and services might arise. This is certainly true of the value of biodiversity to the pharmaceutical industry and their customers. Another example is that the value of forests for carbon sequestration was not known as an important value to humans before the problem of global warming was identified (Bingham et al. 1995).

As a result of this forum, the EPA sponsored case studies, one of which is a project now underway at the University of Maryland that brings together economists and ecologists to model jointly the Patuxent watershed of the Chesapeake Bay ecosystem. The Chesapeake Bay has been studied extensively by economists and ecologists separately in the past, but not in concert previously. We are integrating ecological and economic modeling and analysis in order to improve the understanding of regional economic and ecological systems, to assess potential future impacts on these synergistic systems of assorted possible policy scenarios, and to ameliorate some of the shortcomings in present attempts at ecosystem valuation.

The first step in this process is to model the interaction of the ecosystem and human activity which will illustrate how humans intervene in the ecosystem and how different ecosystem configurations contribute to human welfare. The joint ecological and economic model we are developing will examine the function and processes of ecosystems, clarify relationships between human actions and ecosystems, and provide detailed information that will enable improved ecosystem valuation. This will provide a means of describing the evolving landscape under different policy scenarios on land-use control, non-point source pollution regulations, and ultimately on the valuation of ecosystems in a more holistic manner than the previous segmented efforts.

The model consists of interrelated ecological and economic models that employ a landscape perspective. This perspective captures the spatial and temporal distributions of the services and functions of the natural system and human-related phenomena such as surrounding land-use patterns and populations distributions. Configuration and reconfiguration of the landscape occurs as a result of ecological and economic factors, and there factors are closely intertwined. The purpose of having an integrated model of the ecosystem and the economy is to reflect more accurately how the distribution of human activities such as farming, electric power generation, commercial and residential development, recreation, wastewater treatment, highway construction, and fishing affect the ecosystem, and at the same time to measure the effect of the ecosystem landscape on the quality and value of goods and services, such as recreation, wildlife enjoyment, water quantity and quality, housing, and environmental aesthetics and, therefore, on human decisions.

In any joint ecological economics modeling, how inclusive or exclusive of the area and the economic actors to be studied will have a large impact on the outcome of the analysis. For example, relationships between the spatial arrangement of the landscape and its impacts on ecological process will differ depending upon the scope of the analysis (Turner et al. 1989; Turner 1989). One added complication of including economics in the ecosystem is that the impacts can be felt far away from the ecosystem through market forces. There are no simple answers to question of scope. It remains one that must be addressed and resolved with each modeling effort.

Economic Valuation of Ecosystems

The values the public places on environmental services can be classified into "use" and "non-use" values. Use values are derived by individuals who either directly or indirectly use a resource for such activities as fishing, hunting, nature watching or swimming. Non-use values pertain to the importance people place on the existence of natural resources for their own account or as a bequest motive for the sake of family, friends or future generations.

Depending upon the type of ecosystem services under consideration, the economic valuation technique differs. There are two basic approaches to valuation: revealed preference, where observable choices by individuals occur; and stated preference, where individuals respond to surveys to proposed choices. Direct outputs and benefits such as the value of fishing and recreation can be measured by hedonic pricing and recreation travel cost models. These methods use a revealed preference approach. The stated preference approach is used in the contingent valuation method, which is the only technique where the full value of use and non-use values can be estimated. More detail is given in Toman's chapter in this volume.

Market-Based Techniques

Valuation techniques are most straightforward when the ecosystem provides a direct service or output that goes into the production of market commodities, such as the value of wetlands for commercial fishing. The market price of the commercial species is then used to value the wetlands, since the wetland contributes food and habitat to the production and growth of the commercial species. This value is found by estimating the marginal contribution of an acre of wetlands to the production of the commercial species, the net of human effort in the harvest. The valuation of wetlands in this case would be measured by estimating the addition to harvest that would result from an extra acre of wetlands. The value of that extra wetland acre would then be simply the marginal productivity estimate multiplied by the harvest price of the commercial species. The important factor here is that there is a market price for the species that acts as a signal of the value of the ecosystem. Many of the services and outputs of ecosystems do not have market prices attached to them in any way. The following methods describe approaches when there are no direct markets involved.

Hedonic Pricing

One of the earliest attempts at ecosystem valuation by economists was to measure the relationship between housing prices and the quantity of environmental amenities such as air quality at the site. The basic underlying economic theory in this hedonic pricing modeling approach is that if there are two houses that are identical in all respects except for the quality of air in the area where the houses are located, the difference in the selling price of the two houses is a reflection of the human value of air quality. The hedonic pricing methodology measures the contribution of each characteristic of a parcel such as lot size, number of rooms in a house, neighborhood, and environmental amenities to the total value of the property. In this way, we can estimate the value of attributes such as air quality that do not have a traditional market price attached to them.

Travel Cost

The third approach to measuring ecosystem values pertains to the recreational uses of ecosystems. Pristine ecosystems, such as our national park system, are highly valued by both individuals and society as a whole. Knowing how individuals value the recreational services of ecosystems such as camping, hiking, fishing, hunting, wildlife viewing, and boating is important to any natural resource management decision. The recreational value will be impacted by any other ecosystem management decision that could affect the quality of the recreational experience. For example, the quality of the air and water at the site will affect the enjoyment of the experience, and will be a result of the decisions concerning forestry, fish harvest, toxic waste cleanup, the level of mining, and the control of water flows in the region (Freeman 1993).

The first problem with measuring these values is that, in general, recreational opportunities are available at little or no charge to individuals. Since individuals do not have to pay the marginal cost of the services provided, there is no market price that reflects their valuation of the ecosystem. One economic approach to measuring the recreational value of ecosystem services is the travel cost method. This methodology uses information on the time and money spent by individuals in order to visit the recreation site. This allows the analyst to estimate the value of the site to those who visit it. Different individuals who visit the site have to travel different distances to reach it, and will therefore demand different numbers of trips. The value of the ecosystem services to recreationalists can then be determined by the amount of money and time that they are willing to spend in order to reach the recreation site.

In order to implement the theoretical methods described above, measures of the quality of the ecosystem's services must be derived. These measures can be based on scientific objective measures such as water quality, dissolved oxygen, or turbidity. Or a subjective measure of quality can be determined by surveying recreationalists, as is done in the contingent valuation literature, which will be discussed next.

Contingent Valuation

The contingent valuation (CV) method is considered the only currently available method for measuring the full use and non-use value of nonmarket public goods. It is a survey technique that collects information about an individual's willingness to pay for such things as environmental quality or natural amenities. This method creates a simulated market for a hypothetical good, since there do not exist any direct observable markets from which to measure people's total use and non-use value of the ecosystem under consideration. Therefore, the CV method is highly flexible in its range of potential applications. Participants in a CV survey are typically given a policy scenario that includes a change in the provision of some environmental amenity and are asked to respond with information about their willingness to pay to secure or avoid this change. In addition, they are asked questions to identify socioeconomic and attitudinal characteristics that can be used to interpret the variations in responses obtained. Resulting measurements of willingness to pay can be used for determining the value of the total use and non-use value of the ecosystem for the public at large.

An Application to the Chesapeake Bay

Much of the ecosystem valuation research has focused on wetlands, mostly due to the current regulatory interest in these ecosystems. The services and outputs of wetlands serve numerous and varied functions. Wetlands can serve as a direct output, as an input into the production of other public or private goods, or can provide non-use, existence, or option values. These many factors complicate wetland valuation.

Wetlands have been lost and threatened due to human activities. They have either been directly lost by land reclamation for agriculture, aquaculture, industry, or housing development, or they have been gradually degraded by pressures from pollution, either directly through sewage disposal or indirectly through urban and agricultural runoff. The wetland can also be damaged by changes in the hydrology, which can lead to such impacts as an increase in sediment loads. A wetland in its natural state performs many functions and provides many services. It provides flood and storm protection, wildlife habitats, pollution control, recreation, and landscape values. It can also provide broader regional benefits which should be considered as well (Pearce and Turner 1990; Bohlen, this volume).

The Chesapeake Bay has been the subject of research given its importance to the region. Previous work by Bockstael, McConnell, and Strand (1989) has analyzed the benefits of improving the water quality of the Bay in response to the Chesapeake Bay Restoration and Protection Plan of 1985. The goal of this plan was to "improve and protect the water quality and living resources of the Chesapeake Bay estuarine system (in order) to restore and maintain the Bay's ecological integrity, productivity, and beneficial uses and to protect public health."[1] As the authors point out, there is no stated connection between the goals of the plan and how those who pay for the plan will benefit from it. Therefore, the aim of this paper is to understand better how people use and enjoy the Bay and how environmental quality of the Bay affects them.

The contingent valuation method was used to measure the value of Bay improvements in water quality to users and nonusers. Individuals in the Baltimore-Washington region were asked if they thought if water quality in the Bay was acceptable or unacceptable for swimming or other water activities. The 57 percent of responders who stated the water quality was unacceptable were then asked how much they would be willing to pay for improved water quality. These responses led to aggregated annual values of $18–$29 million for nonusers and $47–$89 million for users.

The travel cost method was used next to measure the value of three activities for the Bay—beach use, boating, and fishing—as a function of water quality measured by the level of nitrogen and phosphorus in the Bay. This measure of water quality is appropriate since one of the most important water quality problems in the Bay is overenrichment, to which these nutrients contribute. The high concentration of nutrients in the Bay decreases the abundance of underwater vegetation, which negatively affects the habitat for fish and birds and degrades the appearance of the water (see Boynton, this volume). The striped bass catch rate was used as a proxy for water quality for the recreational fishing estimation. The benefits of a 20 percent decrease in the two nutrient loadings, and a 20 percent increase in striped bass catch were then estimated. The annual aggregate benefits for the region were: for beach goers, $17–$44 million; for boating, $650 thousand to $8 million; and for fishing, $660 thousand to $2 million.

The Ecological Economics Patuxent Project

The Ecological Economics Patuxent Project has brought together economists and ecologists at University of Maryland, the Department of Agricultural and Resource Economics, and the Center of Environmental and Estuarine Studies in order to reconcile some of the differences that have emerged from the separate disciplines' approaches to ecosystem valuation. The ultimate goal of this project is to gain insight into the dilemma of valuing ecosystem configurations in the landscape and the services derived from them.

Land use patterns dramatically affect the health and status of an ecosystem. It has become increasingly clear, for example, that nonpoint source pollution of ground and surface water is driven by the amount and pattern of forest cover, agricultural activity, residential sprawl, and urban development in the watershed. Likewise, both the abundance and diversity of wildlife, including threatened species, are diminished through habitat alteration by deforestation and fragmentation of the landscape. Even regional air pollution and its deposition in surface waters are affected by land use patterns, because residential sprawl extending great distances from employment centers leads to increasing automobile emissions.

A key element in the above discussion is *pattern*. It is not just the *amount* and type of human activity in a watershed that matters; it is also the *spatial distribution* of that activity. Location matters because the environment's sensitivity or ability to assimilate contaminants varies (e.g., activities in close proximity to a tributary are more likely to contaminate the aquatic environment; activities undertaken on sandy soils are more likely to contaminate groundwater, etc.). Location also matters because the relationship between human activity level and ecological effect may not be linear, i.e., a given amount of effluents discharged at one site may have a different effect on the environment than the same aggregate amount discharged uniformly over ten sites. Additionally, location matters because scale and fragmentation matter. Some species may require a large contiguous area of forest, while others may survive in small parcels of open space, and for some species the edges of the habitat may matter (e.g., a forest buffered by an urban area is different from the same forest buffered by agriculture).

The pattern of the landscape matters to humans through its effect on the environment, but also because of its effect on aesthetics, access, noise, congestion, and other amenities/disamenities that alter the quality of life. The nearer the activity, the greater the amenity or disamenity effect, making proximity or location important for these considerations as well. There are, then, enormous environmental and other externalities associated with land use pattern, and these externalities have strong spatial dimensions.

Despite the externalities associated with land use pattern, land use decisions are made principally by private individuals. However, there are regulations that directly constrain these decisions, e.g., zoning and land use management controls, wetlands permitting requirements, critical areas designations, etc. In addition, other public policies have significant, although often unintentional, effects on these decisions. Transportation policies, especially highway construction and gasoline pricing, affect the costliness (both in money and time) of residential sprawl. Agricultural policies and regulations that make specific types of farming more or less profitable, as well as programs such as those for transferable development rights, alter the proclivity of farmers to sell agricultural land for development. Local tax policies offer reduced tax rates for agricultural and forest land encouraging land to stay in open uses, while federal and corporate income tax laws provide incentives for new construction and home ownership. Finally, environmental policies that impose effluent standards, preclude certain practices, or provide subsidies for voluntary actions all have some effect on the profitability of putting land, or keeping land, in any given use.

What emerges from the interplay of public policies and private decisions, both driven in part by external factors like regional population and economic growth pressures, is a continually changing landscape. The broad goal of the research proposed here is to understand better the process by which the landscape is reconfigured. Such an understanding would provide the means for predicting the land use patterns that are likely to emerge from differing policy environments. These changing land use patterns have direct effects on the quality of life, as well as indirect effects on humans through their ecological consequences. The ecological consequences of special interest include water quality and quantity dimensions of the Patuxent estuary, an important part of the Chesapeake Bay hydrological system. In order to make progress in this direction, one needs to understand the physical connection between the distribution of human activity in the landscape and its effects on characteristics of the ecosystem. One also needs to understand what causes that distribution of human activity to change over time.

The watershed chosen for the case study is the Patuxent watershed in southern Maryland, one of the nine river basins of the Chesapeake watershed and covering about 1,000 square miles. This includes parts of seven counties, ranging from the Washington, D.C. suburbs and the state capital to predominantly rural counties at varying stages of development (see Figure 7-1). Significant portions of land within the area are dedicated to each of the major land uses—commercial, high/medium/low density residential, agriculture (mainly cropland and pasture, with few orchards), forests (both deciduous and coniferous), and wetlands. There are a few industrial centers and some military establishments. Agriculture comprises 32 percent of the watershed's land, and forests comprise 46 percent.

Figure 7-2 provides an overview of the components of the integrated model and a representation of their interactions. The economic and ecological models exist separately, but in parallel, and exchange information on ecological and economic elements. This approach preserves the integrity and intuition of both models. It also allows the appropriate choice of time step, geographical scale, and level of aggregation which might differ between the ecological and economic models. The interactions between these models lead to predictions of a new configuration of the landscape. The integrated model shows for example, the spatial impact of government policies concerning zoning and critical areas to protect water quality.

The Ecological Model

The ecological model is based on the Patuxent Landscape Model, one of several landscape-level spatial simulation models currently under development by Costanza and others at the University of Maryland. This model continues previous research which developed the coastal ecological landscape spatial simulation model (Costanza et al. 1990). This previous model has been used in the Atchafalaya Delta area of coastal Louisiana to model spatial ecosystem processes, succession, and land loss problems, as well as to evaluate the impacts of management strategies and specific projects designed to alleviate coastal erosion problems. The model is also being implemented in the Water Conservation Area and Everglades National Park in Florida to examine the impact of management strategies on water levels, nutrients, and plant succession patterns (Fitz et al. 1993; Fitz et al. 1996). This model is capable of simulating the succession of complex ecological systems using a landscape perspective. It is now being calibrated for the Patuxent watershed.

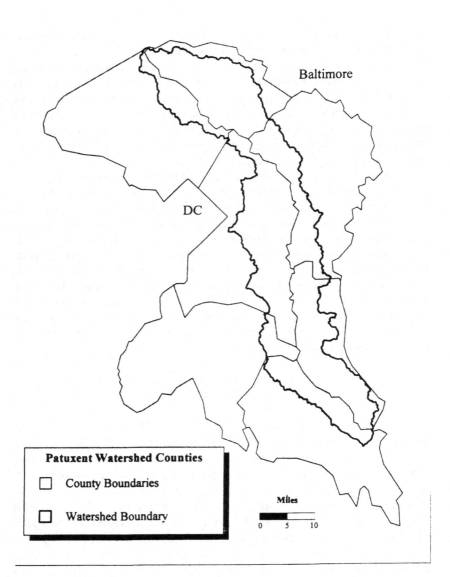

Figure 7–1.   Patuxent Watershed Counties.

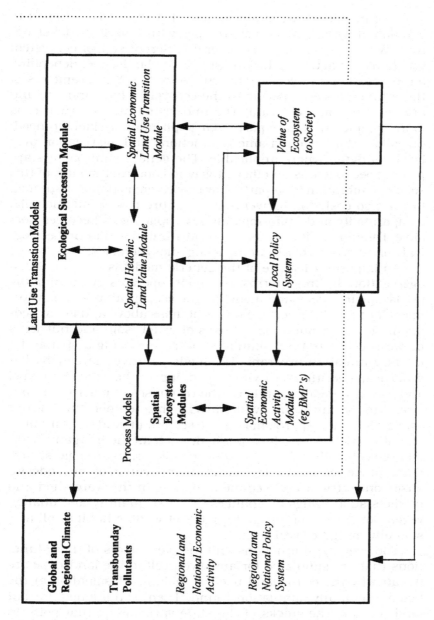

Figure 7–2. Integrated ecological economics model of the Patuxent watershed. Ecology modules are in **bold**; economics in *italics.*

A compelling feature of this model is that it is designed to simulate a variety of ecosystem types with a fixed model structure. While the structure is general, different sets of ecosystem functions are activated for any site in the landscape, depending on its location and ecosystem type. Additionally, parameters of these functions are specific to the ecosystem type and site and are derived from field data. The underlying model structure is more complex than any particular application is likely to need, but allows for selection among functions and aggregation over levels of detail where applicable. The generic approach is appealing because it is an efficient way to construct models of this sort. Recalibration for a particular ecosystem is time consuming, but not so costly as reinventing the entire model. Additionally, comparability and uniformity across applications becomes possible. Differences in results can be attributed to differing ecological conditions rather than modeling idiosyncrasies.

An important feature of the generic model is its spatial disaggregation. In broad terms, the model operates by dividing the landscape into cells and modeling the ecological functions within each cell and the vertical fluxes of mass above and below sediment. The horizontal mass fluxes of water, soil, and nutrients between cells are then simulated over time using a spatial dynamic simulation program. The model is driven largely by hydrological algorithms (varying depending on the ecosystem type) and focuses predominantly on the responses of macro- and microphytes to nutrient availability, light, temperature, water availability, etc. Approximately 14 sectors (including a number of state variables) are incorporated, such as the inorganic sediments sector, dissolved phosphorus sector, hydrologic sector, macrophyte sector, etc. The Patuxent application of the model focuses on nutrient and sediment loading in the watershed and predicts such things as changes in water quantity and quality, vegetation, and amount and quality of wildlife habitat, all at a spatially disaggregated level.

The ecosystem functions and the parameters of those functions that are simulated for any given cell in the landscape are dictated by the cell's "land use" or "habitat" designation at the beginning of any simulation round. Then, conditioned on that land use and the stocks of the state variables at that point in time in the cell, the processes and fluxes are calculated. Conceptually, there are two "levels" at which human behavior could be expected to affect this simulation. One is in the land use des-

ignation of a cell; the other is in the nature of ecological processes that occur within a cell conditioned on its land use.

The Economic Models

The ecosystem model, without economic input, imposes
rather than models human behavior. Consider the land use designation. The ecological model calculates land use designation
through a "habitat switching" model which determines when,
through natural succession or weather-driven ecological cata-
strophe (e.g. flood, forest fire), the habitat shifts from one type to
another. Human-instigated land use changes must be imposed
exogenously and hypothetically. Therefore, the first important
contribution of the economists is to model this human land use
conversion and how it is related to both the ecological and economic features of the landscape.

Human interactions with the environment, conditioned on
land use, are similarly imposed in the current ecological model.
For example, if a cell is designated as being in cropland, then a
given set of processes and parameters are assumed to operate,
conditioned on ecological features such as the slope of the land
and the soil types. Variation across individuals or responses to
external stimulae, like changing prices, are ignored. In order to
assess the effects of some non-point source policy, the model
must impose an assumed change in these processes and parameters, ignoring human response to the change in the regulatory environment. The second type of contribution that the economists are making is in modeling these conditional human
interactions. Our first endeavor of this sort involves modeling the
behavior of farmers, both crop choice and the adoption of best
management practices, as functions of ecological and economic
forces.

The Land Conversion Model

Recognizing that the ecological effects of human activity are
driven by the specific uses man chooses to make of the stock of
natural capital, one of the major contributions we can make to
the ecologists' landscape model is an understanding of how the
land use decisions are made by individuals. This is critical for
the integrated modeling effort, since the simulation of each geographically designated cell's ecological functions is driven by

land use designation. Figure 7–3 shows the land use changes that occurred during the period 1985–1990 for the Solomons region of the Patuxent watershed, as well as housing development that occurred during that same period. To the east of Solomons is the Chesapeake Bay, to the west the Patuxent River. Changes similar to these occurred throughout the watershed during the same time frame. The goal of the land use conversion model is to explain why these land use changes have occurred over time, given ecological and environmental features, economic issues, and government policies such as zoning.

What happens to land, not just aggregate land but the spatial arrangement of land, is a topic of increasing interest to multiple disciplines. Land use is inextricably tied to public infrastructure demands that are more or less costly depending on their spatial distribution. Land use is almost synonymous with locational externalities—visual, noise, etc. Also, land use has environmental consequences that differ markedly depending on the pattern of remaining habitat and the size and proximity of disturbances to ecologically sensitive areas. The configuration of land is one of the major contributors to the quality of life.

There are some spatial attributes that are clearly important to the value of land in different uses. For example, the value of a parcel in residential use will be affected by access to employment centers (transportation networks and proximity to business districts) and private and public infrastructure (shopping, schools, recreational facilities), etc. But it will also be affected by the spatial arrangement of ecological features and man-made structures making different parcels equidistant from employment centers of differing value because of these spatially oriented amenities/disamenities. Additionally, the ability to convert land to a developed use is circumscribed by regulatory mechanisms and incentives: zoning, land use controls, taxation patterns, Best Management Practice incentives, etc. The value to society of land in an undeveloped state also depends on attributes of the land and its spatial arrangement. For example, the suitability of a patch for wildlife habitat depends on its water and vegetative features, its size, shape, and habitat edges, and its proximity to human disturbances and human access.

However, traditional fields of economics have reduced the complexity of spatial relationships, almost to the point of making spatial issues nonexistent. Either aggregate relationships have been specified or the spatial components in a model have been reduced to uni-dimensional variables, e.g., the distance be-

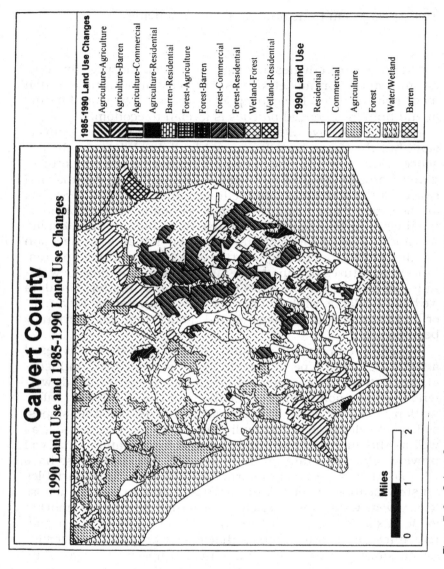

Calvert County
1990 Land Use and 1985–1990 Land Use Changes

1985-1990 Land Use Changes
Agriculture-Agriculture
Agriculture-Barren
Agriculture-Commercial
Agriculture-Residential
Barren-Residential
Forest-Agriculture
Forest-Barren
Forest-Commercial
Forest-Residential
Wetland-Forest
Wetland-Residential

1990 Land Use
Residential
Commercial
Agriculture
Forest
Water/Wetland
Barren

Miles
0 1 2

Figure 7–3. Calvert County 1990 Land Use and 1985–1990 Land Use Changes.

tween economic activities in a location model, the wage differential in a migration model, or cost of access in a transportation mode choice model. The concept of a landscape *mosaic* of natural and human-managed *patches* is foreign to economists.

Economists have often attempted to model land use conversion, but they have been hampered by limited data. In the absence of spatially articulated data, there has been no impetus to develop broadly adopted methods for analyzing two-dimensional space. But now that Geographic Information Systems (GIS) data are becoming more readily available, economists are reconsidering their analytical tools. We can use these data to think explicitly about spatial interactions and their impacts on decision making beyond including location-specific amenities and distances to features of importance. We can also use these new data to describe better the aspects of space that matter. The data we have available, while not perfect, offer the potential for a richer and more spatially disaggregated model than has previously been possible.

Our GIS data include mappings of land uses such as types of agricultural land, types of forests, types of residential, industrial, institutional or commercial development, barren land, wetlands, at four points in time for the counties of interest. Information is also available for ecological features, such as slopes, soil types, elevations, and hydrology (streams, rivers, etc.), as well as the output of the ecological model simulations that will provide values for state variables in a GIS format. We also have GIS-based tax assessment information, which includes variables on size, location, zoning, land use designation, property factors (e.g. sewer, water, historic, etc.), structure description, market value, tax assessment, building value, land value, etc. Of particular interest are the variables that report property transfer information, if a structure exists on the property the year of its construction. This is all in addition to the GIS-level data supplied by the ecological model. These data allow us to calculate, for any arbitrarily small cell in the landscape, such things as distances from roads and highways, towns and employment centers, and natural ecological features of interest—like shoreline or recreational facilities. They also provide a means of calculating variables that reflect what is going on around a particular point on the landscape and what may be happening to the quality of the environment.

The economic model that is under development for use with the ecological landscape model will model the human-induced land use change. As previously discussed, at present the ecology model must exogenously impose these changes. By doing this, the ecology model misses the most important elements of land use changes—their interrelationship over time and space, how humans impact the spatial landscape, and in turn, how the spatial landscape influences the utility level of individuals. The ultimate goal of the economic model is to be able to predict on an annual basis the probability that a given parcel of land, of a given description and in a given location, will remain in its current use or be converted to any one of a number of alternative uses. While the conversion process will be affected by inertia and other disequilibrium considerations and constrained by zoning and other land use controls, the conversion probabilities are likely to be functions of the value of the parcel in alternative uses. Thus, the first step is to model property values in a hedonic framework.

When asked what determines the selling price of a house, realtors reportedly answer "location." Traditional hedonic studies have attempted to capture location by access or distance measures, thereby reducing the spatial components of the model to a vector of uni-dimensional variables. If indeed, this is the principal means by which space enters the problem, then the advantage of GIS type data is solely in making distance measures more accurately or more easily computable. Even the most basic GIS software can calculate distances between points (as the crow flies)—e.g., accessibility to an employment center, to waterfront recreation, to the public library or the local school. More sophisticated software can calculate actual driving distances.

But the importance of location in land values and land use determination is not restricted to accessibility. Externalities characterize land use, and these externalities are spatially determined. What is around a property has a major influence on its value. Economists have attempted to capture this using neighborhood characteristics or indices of neighborhood quality, based on income levels, crime statistics, and so forth. Once again, spatial considerations are reduced to uni-dimensional measures, in this case obtained from aggregate data such as census statistics.

There are at least two problems with this approach. The first is that neighborhood delineations are arbitrary and misleading. Locational characteristics are more likely characterized by a gradient than by discrete levels that change abruptly over polygon boundaries. The second is that there may be more to locational amenity effects than the ones traditionally measured. Land use externalities include not only neighborhood effects such as crime rates, but also proximity to externalities such as traffic, industrial noise, and visual amenities/disamenities. Individuals will value *patterns* of land use surrounding a property.

Unlike most traditional hedonic studies that focus on rent gradients in urban areas or on the suburban fringe, our problem is one in which we need to explain the variation in property values over a diverse region. As such, the above concerns are even more troublesome. For example, in rural or regional, as opposed to the urban, settings, the concept of well-defined neighborhoods may be quite limited. The small towns that are dispersed throughout these counties may be considered "neighborhoods," and perhaps less so certain suburban areas close to Baltimore and Washington, but much of the area is agriculture, forest, or low-density residential sprawl. Here local amenity effects will be better captured with sliding boundaries or with gradients associated with distance from site. This means that specific location matters and the inclusion of uniform "neighborhood" characteristics based on aggregate census block group data may cause serious errors in variables.

Some amenities or disamenities are very specific to the particular property because they are related to those visual and noise externalities that impinge directly on its inhabitants. Thus, knowing the exact location and the land use surrounding that location may add explanatory power to a hedonic function. Some of the most sophisticated GIS data and software provide three-dimensional information, making it possible to determine the visibility of an undesirable land use, for example, from any given location in the landscape.

Hedonic models are used to explain and predict property values and are usually based on assumptions of single markets that are continuously in equilibrium. Equilibrium prices are modeled as functions of the characteristics of the real property itself (structural characteristics) and characteristics of the location or neighborhood. The preliminary model discussed here was developed mainly to investigate specific technical spatial econometric relationships, for housing transaction in 1990 for Calvert County, one of the seven counties of the Patuxent watershed (Geoghegan and Bockstael 1995). We first followed the traditional approach and then investigated the gains that could be made by incorporating more spatially explicit data and by taking into account possible spatial dependence and spatial heterogeneity (Anselin 1988).

Figure 7–4 shows the distribution of property transactions on 1990 and the associated land use for the Solomon's Island area of Calvert County. By inspection, it is obvious that waterfront property is highly desirable and valuable. Comparing Figure 7–4 to Figure 7–3, we can see that most of the lower valued transactions occurred on land that was converted recently to residential use. The preliminary hedonic model attempts to capture the contribution of these and other characteristics to the total value of the property.

This preliminary hedonic model was estimated under a number of different model structures. The explanatory variables were: if the property was on waterfront, which captures all the amenities associated with water views and water access; the lot size of the property; the total square footage of the house; distance to the northern highway access point of the county, which relates to the commuting costs to the Washington, D.C. metro area; and the amount of low-density housing in a 0.1 mile radius around each property, which captures some of the spatial amenities associated with land use. These last two variables were created using the GIS.

In all versions of the model, the estimated coefficients were of the expected sign. Waterfront adds considerably to the value of a house, the larger the house and lot size the more expensive the property, and the closer the property was to Washington the more valuable the property, reflecting the value of short commutes. The amount of low-density residential land in the immediate neighborhood adds positively to the value of each property, although this variable was not always strongly significant.

In one interesting version of the model, we consider the possibility that the hedonic function varies continually over space, that is, the characteristics of the property are valued differently across the landscape. To test this hypothesis of spatial heterogeneity, we employ varying parameters models where the parameters vary linearly and quadratically with distance from the major employment centers. The results for the model to see if the values vary in a linear relationship from the north point of the county show no effects. However, when we consider that the parameters may be nonlinear in distance, some interesting effects emerge. The results of this model suggest that the marginal attributes prices for both waterfront property and lot size vary significantly with distance. These marginal prices begin high, decrease, and then increase again with distance from Washington. This may reflect the premium associated with close commuting

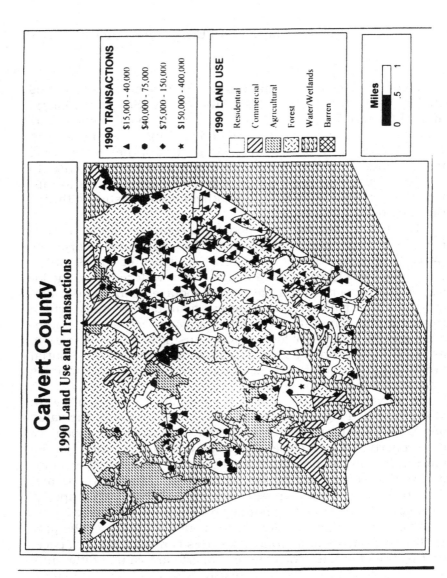

Figure 7–4. Calvert County 1990 Land Use and Transactions.

distances on the one hand, and the premium associated with the natural amenities associated with the southern extremity of the county (including easier access to both the Chesapeake Bay and the Patuxent River) on the other.

This hedonic model will be refined further to include the transactions that occurred in years previous to 1990, as well as for the other remaining counties of the Patuxent watershed. This model will then be used to predict land values for alternative land uses for any given parcel of land in the watershed. These alternative land values will then be used to predict the probability that a parcel of land will stay in its present use or convert to another land use.

Agricultural Activity in the Patuxent Basin

Agriculture is a major source of nutrient pollution in the Patuxent watershed. Although land classified as agricultural represented 30 percent of the Patuxent watershed in 1990 (Patuxent Estuary Demonstration Project 1994), agricultural activity was estimated to contribute 57 percent of the watershed's phosphorus loadings and 46 percent of its nitrogen loadings. The nitrogen concentrations in the Patuxent have been more difficult to reduce, and appear to be the major obstacle in achieving Maryland's goal of a 40 percent reduction in nutrient loading in each watershed by 2000.

Understanding adoption rates of Best Management Practices (BMPs) in agriculture is an important element in the Chesapeake Bay Program and in the development of an economic land use model for the Patuxent. In order to model agricultural practices, we have two surveys of Maryland farmers which focus on adoption patterns in BMPs. It appears that farmers in counties adjacent to the Patuxent River basin are somewhat distinct from farmers in other regions in Maryland. Economic returns to farming by Patuxent farmers are modest, with an estimate of 65 percent of the 1991 Patuxent agricultural units receiving 50 percent or less of their net income from farming and having annual farm sales of around $7,000.

Understanding agricultural land use and activity is an important component of the ecological impacts on the region by human activities. The transition to and from agriculture is a crucial element of the overall land use in the region. By matching the sample of farmers described above with the previous GIS-based land use information, we will explore the characteristics of change that created small farm operations. Did they, for example, begin as larger commercial operations that sold portions of the original farm to develop residential communities? Have small or large farms been created from woodlands, and what are the characteristics of the farms that have converted? The same general methodology expressed in the land use model will be used here.

Secondly, the management of nutrients in the Patuxent basin may require modification of the agricultural land use designations. Despite the fact that about two-thirds of the land in agriculture is operated by the more commercial farmers, over half are not traditional, receiving little income from farming, and are tied to other off-farm income-earning activities. These farmers are unlikely to adopt anything that requires a heavy commitment of managerial time. This is also true for many commercial farmers who may have "busy" times of the year and conflicts with time commitments to certain BMPs. These so-called managerial BMPs are less likely to be adopted compared with BMPs that require primarily out-of-pocket expenses, like construction of structures. Government programs that only cover economic costs that do not include time costs will then have a different effect on different farmers. Understanding the connection between different types of BMPs and farmers' behavior will have a major impact on the management of nutrients in the Patuxent watershed.

This part of the model will let us predict the probabilities that a parcel of land is used for agriculture in the Patuxent watershed, is a particular type of farming operation, uses a particular set of BMPs, and hence, discharges into surface and groundwater predictable amounts of nitrogen and phosphorus. Taking the expectation of these probability statements will yield an expected amount of nitrogen or phosphorus entering into the ecological model.

The Integrated Model

The major components of the integrated model are spatially explicit, dynamic models for the ecological system and the economic decisions, and the mechanism for exchange of information between the two. The results of the interrelated processes produce a new landscape and a new array of values associated with the landscape. Some of these become signals to private individuals and public agents in future decisions.

Also, as part of this exercise, the ease and cost-effectiveness of generalizing the model to other regions will be assessed. It is expected that the modeling approaches will remain fairly constant over application, although the particular behavioral models and ecological processes emphasized, as well as the actual parameter of the models, will differ depending upon the application. Because the models are designed to depict specific processes and ecosystems and to evaluate specific management strategies, they require a significant amount of site-specific data. However, the integrated model will illustrate the circumstances under which human actions set in motion processes that can have far-reaching effects. These effects can play out over a broad geographical range and over time. These illustrations will help policy makers define the relevant questions to be asked and effects to be anticipated in other regional settings (Bockstael et al. 1995).

Conclusions

This paper has demonstrated the varied ways in which economists value ecosystem functions and outputs, as well as previous research using these methods into specific ecosystem valuation. However, the major focus of previous economic research into ecosystem valuation has been at looking at specific, tangible, fragmented commodities derived from the ecosystem that have been of specific or obvious immediate benefit to humans. As has been argued in this paper, these measures are incomplete since this approach ignores the value of the entire ecosystem as a whole, and they cannot measure the changes that occur in ecosystems over time and space, and do not address the possibilities of changes that are irreversible and detrimental. Considering only those services of ecosystems that are well-defined, are easily measurable using conventional market or non-market valuation methods, and have immediate consequences for humans, ignores the more subtle, long-range contributions of the ecosystem to human welfare. It also ignores the importance of the configuration of the ecosystem landscape in determining its values, since the geographic location of things matters.

References

Anselin, Luc. 1988. *Spatial Econometrics: Methods and Models*. Kluwer Academic Publishers, Dordrecht, The Netherlands.

Bingham, G., R. Bishop, M. Brody, D. Bromley, and others. 1995. Issues in ecosystem valuation: Improving information for decision making. *Ecological Economics* 14:73–90.

Bockstael, N. E., K. E. McConnell, and I. E. Strand. 1989. Measuring the benefits of improvements in water quality: The Chesapeake Bay. *Marine Resource Economics* 6:1–18.

Bockstael, N., R. Costanza, I. Strand, W. Boynton, K. Bell, and L. Wainger. 1995. Ecological economic modeling and valuation of ecosystems. *Ecological Economics* 14:143–153.

Costanza, R., S. Farber, and J. Maxwell. 1989. The valuation and management of wetland ecosystems. *Ecological Economics* 1:335–61.

Costanza, R., F. H. Sklar, and M. L. White. 1990. Modeling coastal landscape dynamics. *Bioscience* 40(2):91–107.

Fitz, H., R. Costanza, and E. Reyes. 1993. *The Everglades Landscape Model (ELM): Summary Report of Task 2. Model Development*. Report to the South Florida Water Management District, Everglades Research Division.

Fitz, H., E. DeBellevue, R. Costanza, R. Boumans, T. Maxwell, and L. Wainger. 1996. Development of a multi-ecosystem simulation model (MESM) for a range of scales and ecosystems. *Ecological Modelling* 88:263–295.

Freeman, A. 1993. *The Measurement of Environmental and Resource Values: Theory and Methods*. Resources for the Future, Washington, DC.

Geoghegan, Jacqueline and Nancy Bockstael. 1995. *Economic Analysis of Spatially Disaggregated Data: Explaining Land Values in a Regional Landscape*. Paper presented to the Association of Environmental and Resource Economists at the Allied Social Sciences Association meeting, Washington, DC., January 1995.

Odum, E. P. 1971. *Fundamentals of Ecology*. 3rd edition. W.B. Saunders, Philadelphia, PA.

Patuxent Estuary Demonstration Project. 1994. *Interum (sic) Guidance Document*, Draft.

Pearce, D., and R. Turner. 1990. *Economics of Natural Resources and the Environment*. The Johns Hopkins University Press, Baltimore, MD.

Turner, M., et al. 1989. Predicting across scales: Theory development and testing. *Landscape Ecology* 3(3/4):245–252.

Turner, M. 1989. Landscape ecology: The effect of pattern on process. *Annual Review of Ecology and Systematics* 20:171–197.

Note

1.   Chesapeake Executive Council, *Chesapeake Bay Restoration and Protection Plan*, U.S. Environmental Protection Agency, Sept. 1985, as quoted in Bockstael, et al., 1989.

CHAPTER EIGHT

"Green" Accounting for the Chesapeake Bay

Henry M. Peskin

What is "Green" Accounting?

Although the term is widely used, "green accounting" does not appear to have a universally accepted definition. Clearly the word "green" suggests that the accounting should have something to do with the natural environment. But which aspects of the environment? Natural ecosystems, rain forests, the oceans? Is green accounting meant to concern itself with air quality and water quality? What about minerals, commercial forests, agricultural soils?

The word "accounting" also appears to cover a number of possible concepts. Some use the term in the general sense of "taking account of [something]" as in " . . . We should take account of losses in our natural resource base." The development of data and information describing these losses would appear to fit such a definition. To others, however, "accounting" evokes the activities of the business or governmental accountant. To this group, not just any data system would suffice but rather only one that approximates the strict conventions of business or national accounting.

The concern of this paper is the development of an information system that can be used to analyze interactions between the natural environment and human economic activity. Thus, the words "green," "environmental," and "natural resources" are used in the broadest possible sense: the intent is to "account" for all natural wealth. However, the accounting in question is an *economic* accounting; only those natural assets that are economically valuable—that is, services that are scarce relative to the social demands placed on them—are accounted for. As we will see below, not all environmental assets have economic value, even though they are quite valuable in a non-economic sense. Such assets are not accounted for. Moreover, because of the concern of showing interrelationships with ordinary economic activity, the system adopts many (but not all) of the strict conventions of the business and national accountant.

Of these accounting conventions, the most important are those that assure completeness. The principal goal of any accounting system is to assure that the accounts provide a complete description of all inputs and outputs that characterize an economic process at one point in time (referred to as the accounting period). Because of the goal of relating information to an economic process, a large body of environmental information—such as much of the technical and scientific data on the environment—has no direct role to play in a "green" accounting system. For this reason, a "green" accounting system should be distinguished from, and not be considered a substitute for, the concept of a more general environmental data system.

Why "Green" Accounting?

Early efforts at "green" accounting, in the sense the term is used in this paper, grew out of a dissatisfaction with the perceived failure of the conventional national economic accounts to provide a meaningful measure of social performance in the face of environmental degradation. Literature expressing concerns about the national account's ability to measure social welfare dates back to the 1930s, but the specific focus on the environment appears to coincide with the environmental movement of the mid-1960s and early 1970s.

While these concerns were often expressed by academics or others knowledgeable about the national accounts, most of the criticism appeared in the popular press. In particular, there was criticism of the GNP or GDP, Gross National or Gross Domestic Product, as indicators of social performance. These indicators, drawn from the national economic accounts, measure the total production of goods and services that meet national consumption, investment, governmental, and net foreign trade needs.[1] While either measure is the most popular indicator of a nation's economic performance, neither adequately reflects any degradation of the nation's environment. Indeed, their movement is often perverse: GNP or GDP could increase in response to environmental degradation. Efforts to clean up an oil spill or increased health expenditures necessitated by poor air quality would tend to increase national production even though these actions are no more than an attempt to maintain environmental quality at an acceptable level.

Many economists join environmentalists in questioning the usefulness of the GNP (or GDP) as an indicator of social welfare. At best, as Professor Hicks pointed out over 50 years ago, GNP could be an index of *economic* welfare (Hicks 1940). But economic welfare is not necessarily social welfare. Furthermore, if relative prices change over time and if some of the goods society desires increase while other goods decrease, the GNP index could either increase or decrease depending on whether "old" or "new" prices are used in the calculation. Finally, at best GNP can be interpreted as a linear approximation to some non-linear social welfare function. But as Arrow pointed out, there are no assurances that such a welfare function even exists—at least as a function that consistently reflects the preferences of members of the society (Arrow 1963). It is no surprise, therefore, that economists are not especially impressed with the criticism that GNP overstates well-being because of its neglect of environmental degradation.

However, there are two other criticisms of the conventional national economic accounts and their neglect of the environment that should have more appeal to the professional economist. First, there is the claim that the accounts treat reproducible wealth and natural wealth inconsistently. Specifically, the stock of reproducible wealth is depreciated in order to calculate an income measure—net national product—that measures a nation's "sustainable" income—income that allows for replacement of losses in the capital stock. However, losses in the stock of natural resources is not similarly depreciated, suggesting that true net national product is overstated.

Second, there is the criticism that the conventional national accounts are incomplete in that they neglect important inputs and outputs in the nation's "production function"—inputs and outputs that have economic significance but are neglected because they lack market-determined values and prices. Typically neglected inputs are the waste disposal services provided by the natural environment. These services are typically scarce (and, thus, have economic value) because they compete with other services also provided by the natural environment such as recreational and ecological services. Because all these environmental services are not bought and sold in conventional markets, they lack prices and therefore are neglected by conventional economic accounting.

Scorekeeping and Management

These criticisms of conventional national economic accounting reflect a failure to attain two principal objectives of accounting: scorekeeping and management. Scorekeeping refers to the calculation of statistics measuring the performance of a business or of an economy. Revenues, assets, liabilities, and profits are the prime examples of such statistics for businesses, while GDP and GNP are examples for a nation's economy. Management, on the other hand, refers to the use of the body of data generated by the accounting process to support day-to-day management of a business or an economy. Of particular importance to the management of either a business or an economy are sets of "micro-data" on costs, production, sales, etc. These data are the basis of the scorekeeping statistics but they also have their own direct use as inputs to formal or informal models used to help formulate policy.

The available data suggest that the scorekeeping value of green accounting may be relatively minor. The differences between estimates of environmentally adjusted GNP or GDP and the conventional statistics, in both developed and less-developed economies, and for the Chesapeake Bay region seem relatively small (less than two percent) and may not be of any statistical significance. There have been studies showing bigger differences. For example, adjustments for Indonesia were in the range of about four percent (Repetto et al. 1989). A recent "green" GDP calculation for the United States, following procedures suggested by the U.N., generated a larger difference: about 7.5 percent (UN Handbook 1993) (Grambsch and Michaels 1994). However, in both cases, a depreciation adjustment for any loss of environmental capital services appears to be far larger than is consistent with standard business accounting practice or with the concept of economic depreciation.[2] Also, in the application of the UN system to the U.S., the adjustment also depends on assuming the desirability of environmental clean-up targets that are far more stringent than would be called for by considerations of economic efficiency.[3]

Furthermore, even if these differences are viewed as significant, the policy significance of the difference is open to question. Suppose "green" GDP is less than conventional GDP. The policy implications of this fact may not be not clear. The theory of environmental management has long established that some resource extraction and some degree of pollution—both factors which serve to make "green" GDP lower than conventional GDP—may be socially optimal. Therefore, the mere fact that the two GDP values differ provides no policy direction. It doesn't do much good to give the manager of a baseball team a more accurate "score" if he or she doesn't know how to respond.

Approaches to "Green" Accounting

International responses to the dissatisfaction with the conventional accounts have varied across countries. The differences in approach appear to reflect differences in emphasis between the scorekeeping and management functions of accounting. For purposes of exposition, the various approaches can be grouped under four headings.

Pollution Expenditure Accounting

One of the first reactions to perceived weaknesses in the conventional economic accounts was to develop data series on pollution abatement and other environmental expenditures. (Of course, to simply develop a data series does not meet the criteria of accounting suggested above.) Such data series have been maintained in the United States since 1972 and are available in other countries such as the OECD members.

As these data refer to measured expenditures already incurred, either due to policy or to standard business and household practice, they should not be considered as additions to the conventional economic accounts but rather as a delineation or re-specification of information already accounted for. Some critics of the conventional accounts have argued that such environmental expenditures are inherently "intermediate" and, therefore, should be deducted from final product indicators such as the GDP to generate an appropriate "green" GDP. Such an adjustment, however, has never been made in the United States. Therefore, the motivation for the statistical series on pollution abatement expenditures appears not to be for better scorekeeping but rather for better management of economic and environmental policy. Specifically, several investigators believe that such information may provide for a better explanation of changes in measured productivity (Dennison 1979; Jorgenson and Wilcoxen 1990).

Physical Accounting

Another approach, characteristic of several "green" accounting systems and especially those in Norway and France, is to measure physical changes in the stock of environmental assets over the accounting period (usually one year). Typically, a decision must be made as to a physical measure that is relevant for some environmental policy concern. A forest, for example, can be physically measured in terms of its acreage, the volume of its timber, the variety of its biota (as evidenced by the number of available species), the stock of non-timber resources such as firewood and grasses, etc. Which of these alternatives is chosen as the physical measure of the forest will depend on what are the relevant policy objectives: commercial timber management, assurance of firewood supply, adequate specie diversity, etc.

Once the physical units are chosen, the accounting amounts to the specification of an "opening" stock: the physical amount of the asset at the beginning of the accounting year; the amount of the stock depleted due to use or to natural causes; the amount increased due to discoveries or, if appropriate, natural growth; and the resulting "closing" stock. This basic framework can be elaborated by linking the amount depleted to economic activity, usually through an input-output model.

In Norway, where this approach has been followed for many years, there is no attempt to use physical accounting to adjust GDP. Rather, the intent is to generate information and analyses to support the nation's economic planning process. Thus, as with pollution expenditure accounting, the motivation for the accounting is not scorekeeping but management. However, unlike the mere assembly of expenditure information, physical accounting does depict an input-output process and, thus, does meet the above definition of accounting.

"Green" Indicators

A third approach, and perhaps the one with the longest history, is to construct "green" GDP or other economic aggregates to replace the conventional ones. This work has proceeded along two parallel paths. First, there has been the effort to replace the GDP or the GNP with alternative indicators of social well-being, usually by altering one or more of the components of the conventional aggregates (subtracting out pollution-abatement expenditures would be an example) or by adding some new components (such as a factor measuring the negative effects of urbanization). The best known example of this approach is the Nordhaus-Tobin MEW (Measure of Economic Welfare) indicator (Nordhaus and Tobin 1973). Similar indicator approaches have been developed by the Japanese (the NNW, Net National Welfare) and, more recently, by Daly and Cobb (Japan, Economic Council 1973) (Daly and Cobb 1989).

A more conservative example of the "green" indicators approach has been provided by Robert Repetto and his colleagues at the World Resources Institute (WRI) (Repetto et al. 1989). The principal thrust of this effort is not to replace the conventional gross income aggregates. Rather, the purpose is to modify the conventional measures of net product: NNP or NDP, defined as gross product less depreciation. Essentially, the idea is to depreciate natural assets such as forests, mineral stocks, fish stocks, and soils in order that reproducible capital and natural capital receive equal treatment in the computation of net income.

Although both the MEW indicator approach and the WRI approach are the best known examples of "green" accounting, neither records the inputs and outputs of an economic process. These approaches, therefore, do not strictly meet the above definition of economic accounting.

Extensions of SNA-Type Systems

The fourth group of approaches builds upon the existing systems of national accounts. In this sense, they are quite conservative. However, because they do not focus on just one element of the conventional accounts, such as depreciation, but rather seek to cover all sectors that may interact with the environment, they are the most ambitious of all the four groups of approaches. In practice, these approaches require most of the same information used by the above approaches plus much more sector-specific information.

Examples are the UN SEEA (United Nations System for Integrated Environmental and Economic Accounting) approach and the Peskin framework, which is the basis of the Chesapeake Bay accounting discussed below.

The Peskin Framework

The Peskin framework was chosen for the Chesapeake Bay accounting because, of the alternatives, it is the only one that generates information that describes the economic efficiency of policies that affect the allocation of environmental services. Such information could be used to analyze the economic efficiency of prospective environmental policy and, therefore, was of interest to the project sponsor, the U.S. Environmental Protection Agency (EPA). In addition, this framework is apparently unique among the alternatives in its strict adherence to principles of economic theory (Hamilton 1994).

The basic principle of the Peskin accounting framework is that environmental assets—natural resources, air sheds, water bodies—should be treated in an accounting system just like reproducible marketed assets to the extent that the services of environmental assets have economic value. These services in fact have economic value if they are scarce in an economic sense: when demand is limited by available supply. If the services and the environmental assets generating them were traded in markets, they would have observable market prices, and thus would be included in the conventional accounts. However, these services are typically not marketed in spite of their economic value, either because property rights to the assets have not been established or because the "owner" of the property right (often the government) chooses not to act as a seller of the services.

The focus on scarcity and economic value serves to limit the scope of the system. Indeed, a huge body of environmental information, some of which may be very important for the development of environmental policy, is not covered. For example, certainly the energy flux generated by the sun is of immense ecologic value. For this reason, were the purpose of the accounting framework to account for generation and use of energy, this particular service of the sun would have the most prominent role. However, in an economic sense, the energy flux service is of no *economic* value since it is excess supply.[4] It is thus ignored in the Peskin framework. However, while the framework does not cover all the information of significance for environmental policy, it does cover the information that is important for developing policies that affect environmental/economic interrelations.

Consistent with the focus on economic value is the related emphasis on the *economic* depreciation of environmental assets. Physical degradation of environmental and natural resource assets are not, in themselves, of much importance in the accounting system unless these physical declines imply economic losses in the value of these assets. As has been discussed elsewhere, the concentration on economic depreciation, as opposed to physical degradation, is consistent with a particular interpretation of "sustainability" (Peskin 1991). The framework generates a modified net income measure that equals conventionally measured net product less the economic depreciation of natural resource and environmental assets. This modified net income is "sustainable" in the sense that the available assets in the economy, both natural and reproducible, could generate this level of income indefinitely. Other concepts of sustainability—such as implied by the phrase "sustaining the environment"—may require more information on the physical degradation of the environment than is provided by the Peskin framework.

A unique feature of the framework is the dual valuation placed on pollution activity. Specifically, pollution is shown to result from the use of the waste disposal services provided by the natural environment. This service, of value to the polluter, may compete with other services provided by the natural environment such as health, recreation, and ecological support. As a result, the waste disposal services also lead to environmental damages. The numerical value of these damages usually will not equal the value of the disposal services—thus the need for a dual valuation. Indeed, according to the economic theory of environmental management, the value of the marginal unit of pollution to the polluter and to any injured party will be the same only if the services of the environment are allocated among all users in an economically efficient manner (Baumol and Oates 1988). In the absence of markets or optimal governmental pollution policy, such efficient allocations are not likely.

Because of the dual valuation, the accounting framework requires a balancing entry. As suggested above, this balancing entry provides some measure of how much the allocation of environmental asset services departs from economic efficiency. Thus, if the level of pollution is valued with a "price" equal to the value of the marginal unit, the balancing entry will equal zero with optimal allocations. Thus, accounting frameworks, such as the UN SEEA framework, which lack dual valuation, can be interpreted as implicitly assuming that the allocation is always economically efficient. This implicit assumption would explain much of the difference between a "green" GDP generated by the UN SEEA framework and the "green" GDP generated by the Peskin framework.

One other feature of the Peskin framework is the explicit accounting for non-waste disposal environmental services that are directly consumed by society. These include recreation services, esthetic services, and the support of ecological systems. This feature was especially important for the Chesapeake Bay accounting since the value of these services far exceeds the value of the Bay's waste disposal services.

Table 8-1 displays a consolidated modified national account consistent with the Peskin framework. Waste Disposal Services and Damages are entered negatively as a matter of convention. Many of these additional entries (e.g., Waste Disposal Services) can be disaggregated to individual industries. For other entries, such disaggregation may not be practical. In particular, disaggregation is not possible when environmental damages result from pollutants arising from more than one industrial sector. However, even in these cases, *geographical* disaggregation is possible, especially when the data have been generated from detailed micro-data sets. Thus, much of the data for the Chesapeake Bay accounting was drawn from data sets developed to support a previous accounting at the national level.

Implementation—The Management Focus

In principle, and in keeping with the theoretical foundations of the framework, the various environmental service entries shown in Table 8-1 should be measured in terms of what consumers of these services would be willing to pay for them or, in

Table 8–1   Modified Income and Product Account

INPUT	OUTPUT
Compensation of Employees Proprietor's Income Indirect Taxes Gross Return to Capital	Personal Consumption Investment Inventory Change Exports Imports Government Goods and Services
CHARGES AGAINST GROSS NATIONAL PRODUCT	GROSS NATIONAL PRODUCT
Environmental Waste Disposal Services (–) a. Air b. Water c. Land	Environmental Damages (–) a. Air b. Water c. Land
	Direct Consumption of Environmental Services Recreational Esthetic Ecological
Net Environmental Benefit (Disbenefit)	
CHARGES AGAINST MODIFIED GROSS NATIONAL PRODUCT	MODIFIED GROSS NATIONAL PRODUCT
Capital Depreciation (–) Natural Resource Depreciation (–)	Capital Depreciation (–) Natural Resource Depreciation (–)
CHARGES AGAINST MODIFIED NET NATIONAL PRODUCT	MODIFIED NET NATIONAL PRODUCT

the case of damages, what they would be willing to pay to avoid them. Techniques for ascertaining these values are well-known in the benefits-costs literature.

However, in practice it is not possible to apply these techniques for valuing the full scope of environmental services used by an entire economy. Rather, the approach was to develop crude estimates, drawing on the available empirical benefits-costs literature. Cost estimates of pollution abatement were used to proxy the value of waste disposal services, while various environmental benefits studies were used to develop damage estimates and estimates of the consumption value of direct environmental services such as recreation.

Because the resulting estimates are very rough approxima-
tions of the true value of environmental asset services, it would
be legitimate to question the value of the entire accounting effort.
If the sole purpose were to generate the bottom line of Table
8–1—a modified "green" GNP—such a criticism would be quite
valid. Indeed, even with the best data available and with unlimited accounting resources, it would be doubtful that such a
"green" estimate could be generated with enough credibility to
gain general acceptance. The estimation techniques themselves
are too controversial.

For this reason alone, recent implementations of the Peskin
framework (e.g., the Philippines ENRAP project) place more emphasis on the management function of accounting rather than
the scorekeeping function (ENRAP 1994). (While this has always
been the intent of the author, funders and "customers" of the accounting effort may have had more ambitious expectations.
Clearly, better scorekeeping was also one of their objectives.)
Specifically, the implementation effort generates a large body of
information useful for the support of environmental policy. For
example, the use of cost estimates to develop waste disposal values and of benefit estimates to develop damage and direct environmental service values provides a basis for targeting policy using benefit-cost criteria. The recent Philippines accounting
indicated that, overall, non-industrial sources of water pollution
were far more serious a problem than industrial sources. An implicit recommendation is that these sources should receive far
more attention than they have in the past. Such recommendations do not always require great precision in the estimates. In
cases where there is more doubt, the crude accounts suggests
areas where data and estimation improvements are likely to have
the better payoffs for policy. Thus, the crude accounts can point
the way towards future improvements.

Application to the Chesapeake

The Chesapeake Bay accounting project was an exploratory
effort to apply comprehensive natural resource and environmental accounting to an important and environmentally sensitive region of the United States. The region was defined as including
the Chesapeake Bay and all surrounding counties that either
bordered on the Bay or its tributaries, in their estuarine por-

tions. This region was named "Chesapeaka." The Chesapeake Bay area was chosen not only for its environmental importance but also because it was a region that was under extensive study. Because it was a subject of a large level of research, the expectation was that data to support accounting would be readily available.

Details of the project have been published briefly elsewhere (Grambsch and Michaels 1993). The basic findings are summarized in Table 8–2. Rather than discuss the detailed methods used to generate the numbers, this paper will comment on more general issues: the problems of applying the Peskin framework to a region and to the Chesapeake Bay in particular, the validity of the findings, and the usefulness of the effort for environmental policy making.

The Problems of Applying the Peskin Framework to the Chesapeake Bay

As this effort was EPA's first attempt at resource and environmental accounting, there was the feeling that it would be best to "start small." The assumption was that it would be easier to develop accounts for a region rather than for the nation as a whole. This assumption proved to be incorrect.

In the first place, a consistent environmental and economic accounting for a region requires that both environmental and economic activity be defined for the same spatial area. Unfortunately, while it may be fairly straightforward to define a region to encompass all its significant environmental activity, markets do not respect geographical borders—especially for regions of less size than a nation. Actually, in the case of the Chesapeake Bay, neither the environment nor the economy respected Chesapeaka's borders.

Depending on the year, up to three-fourths of all the fresh water that enters the Bay comes from the Susquehanna River, which stretches up to the New York State border. The Potomac and James rivers reach beyond Chesapeaka's borders as well. As a result, about 43 percent of water-borne nitrogen loads and 36 percent of phosphorous loads in the Bay arises from non-Chesapeakan sources. The "imported" contribution of nitrogen rises to 75 percent if air-borne nitrogen that is deposited into the Bay is also considered to be imported. These imports (largely from farms in Pennsylvania) created a special problem in determining

Table 8-2 National Income and Product Account for
Chesapeakea—1985
(millions of 1987$—illustrative but unrevised data)

INPUT		OUTPUT	
Compensation of Employees	94,075	Personal Consumption	81,332
Proprietor's Income	9,978	Investment	17,860
Indirect Taxes	9,978	Inventory Change	884
Gross Return to Capital	28,508	Exports	14,382
		Imports	(11,318)
		Federal Government Goods and Services	22,621
		State Government Goods and Services	16,762
		Statistical Discrepancy	14
CHARGES AGAINST GROSS CHESAPEAKA PRODUCT	142,538	GROSS CHESAPEAKA PRODUCT	142,538
Environmental Waste Disposal Services (–)		Environmental Damages (–)	
a. Air	(104)	a. Air	(110)
b. Water	(90)	b. Water	(347)
Total	(194)	Total	(457)
		Final Consumption of Environmental Services	
		Beach Use	253
		Boating (Chesapeakans Only)	140
		Recreational Fishing	41
		Hiking	185
		Camping	160
		Waterfowl and Deer Hunting	
		By Chesapeakans	122
		By Visitors	8
		Total	130
		Wildlife Viewing	
		By Chesapeakans	149
		By Visitors	45
		Total	194
		Pollution Reduction by Marshland	30

(Continued)

Table 8–2   National Income and Product Account for
Chesapeakea—1985 (Continued)

		Total Final Environmental Services	1,132
Net Environmental Benefit (Disbenefit)	869		
CHARGES AGAINST MODIFIED GROSS CHESAPEAKA PRODUCT	143,213	MODIFIED GROSS CHESAPEAKA PRODUCT	143,213
Capital Consumption (–)	(17,195)	Capital Consumption (–)	(17,195)
Environmental Depreciation (–)	(65)	Environmental Depreciation (–)	(65)
CHARGES AGAINST MODIFIED NET CHESAPEAKA PRODUCT	125,953	NET CHESAPEAKA PRODUCT	125,953

how much reduction in these effluents by Chesapeakans would be required to attain non-damaging water quality concentrations of these two pollutants—a key factor in estimating the costs of pollution reduction. (These costs, in turn, were used as proxies for the value of waste-disposal services.) More reduction, and thus more costs, would be required if the imports were considered non-controllable; less reductions and, thus, lower costs would be required if the imports were controllable. Costs, in fact, were calculated under both assumptions, although only the non-controllable figure was used to calculate the water-disposal value estimate in Table 8–2, above.

A different sort of environmental spillover problem was caused by fish migration. While an original intent was to estimate the depletion and associated depreciation of fish stocks in the Bay, it soon became apparent that the Bay does not "own" most of its fishing stock. The fact is that most species are only seasonal visitors. While catch rates for many species have declined, it is not clear how much of this decline is due to environmental conditions or fishing effort specifically in the Bay. Environmental conditions off the coast of New Jersey or fishing effort near Newfoundland, for example, could have had far more influence. To address this issue, an effort was made to estimate not the change in fish stock but rather the change in biological conditions that support fish stock. Although this effort, based on biological input-output modelling, looked promising, it was never completed.

The economic spillovers caused even more difficulties than the environmental ones. While the Bureau of Economic Analysis (BEA) estimates gross product originating (i.e., value added) by state, there are no corresponding data for the "output" side of the accounts (that is, there are no regional or state data for consumption, investment, government, and net foreign trade). To complete the economic side of the accounts, estimates of these outputs were based on factors drawn from unpublished regional input-output tables prepared by Jack Faucett of JFK Associates. The BEA state data also had to be prorated to Chesapeakan counties using ratios based on county employment. The resulting estimates are not only crude but difficult to interpret. For example, the $81 billion in estimated consumption outlays in Chesapeaka does not necessarily imply $81 billion in Chesapeaka production of consumption goods—at least, not for the same goods consumed. The fact is that the procedure misses a large portion of interstate exports and imports. Most likely, because of the presence of Washington, D.C. in Chesapeaka, Chesapeaka is a net exporter of services and a net importer of goods.

In addition to the technical difficulties caused by economic and environmental spillovers, application of the Peskin framework was continually plagued by poor data availability. While there are rich data sets describing the Bay, the vast majority of these data measure ambient conditions of the Bay. While there are many estimates of the sources of pollutant loadings, there are virtually no data that explicitly and unambiguously show how changes in human economic behavior will affect *ambient* Bay conditions.[5] In fact, unpublished USGS data suggest that less than 10 percent of changes in tributary loads into the Bay has anything to do with human activity. Natural rainfall patterns and their effect on river flow may have the dominant role.

Validity of the Findings

As a result of the data difficulties and the above spillover problems, the resulting estimates were, for the most part, based on crude extrapolations of cost and benefit estimates drawn from other studies, most of which had nothing to do with the Bay. Yet, the rough nature of the estimates does not necessarily imply that there are no findings with any validity. Indeed, there are two findings, in particular, that are of interest and are unlikely to change even with better data.

The most significant of these two findings is the relative smallness of the "green" adjustments. Conventional gross product and environmentally adjusted gross product differ by about one-half percent. This result was not unexpected, given the relative "cleanness" of the Chesapeakan environment. In a previous application of the Peskin framework to the U.S. economy, the difference was similar (1.5 percent for 1978). Interestingly, an adjustment to GDP for the far more polluted Philippine economy yielded even a smaller percentage difference (0.12 percent for 1988).

Of course, other suggested alterations in the conventional GNP could have produced larger differences. For example, one version of the UN SEEA system suggests that the potential costs to "maintain" the environment at a clean level should be subtracted from conventional GDP. This maintenance-cost value is equivalent to our measure of waste disposal services. If this advice were followed for the Philippines and no other adjustments were made, the difference between conventional and "green" GDP would grow to 2.2 percent. However, even this difference approximates the estimated statistical discrepancy in the Philippine accounts (about 1.6 percent in 1988). One is left with the overall impression that differences between standard income aggregates, such as the GDP, and their "green" versions are not of much statistical interest.

On the other hand, a second finding is of more interest. When one investigates the factors that enter the revised GDP estimate for Chesapeaka, it is clear that if there are environmental problems in the Bay, one should look first to those that affect those direct services to final demand generated by the environment. These services, such as beach use, boating, camping, wildlife viewing, etc., as a group have a value that is nearly three times larger than the negative value of pollution. It is unlikely that this qualitative difference will be greatly changed even with more refined estimates.

Usefulness of the Effort for Policy Making

From the beginning, the sponsor, the U.S. Environmental Protection Agency (EPA), recognized the experimental nature of the project. Nevertheless, the hope expressed by some at EPA was that the project would generate information that could be used to assess EPA's progress in Bay policy—a clear scorekeeping objective. Those closer to the project, however, appreciated the potential benefits of the effort as a generator of information to support policy—the management function. Yet both "scorekeepers" and "managers" would have reason to feel some disappointment with the outcome of the project.

As noted, quality data on the value of recreational and ecological services of the Bay environment, on the value of environmental damages to the Bay, and on the costs of rectifying these damages (required for the calculation of waste disposal services) was nearly totally absent. Estimates, therefore, were based on extrapolation of information developed for other sites or from national averages of per unit costs and benefits developed to support EPA regulations. Given the lack of pertinent data that were specific to the Bay region, it is understandable not to have much confidence in any resulting "green" GDP figures. Perhaps, one could take comfort in the finding that the differences between adjusted and non-adjusted GDP figures are quite small and that even with order-of-magnitude improvements in the estimates, the differences would still likely be small. Thus, even if the numbers were accurate, the scorekeeping gains from having a "greener" GDP may be equally small and not worth the effort.

On the other hand, the findings regarding the magnitude of pollution damages relative to the size of the direct final demand services of the Bay may have some policy significance. In addition, the "negative" finding regarding the lack of data linking human activity to ambient environmental conditions in the Bay may similarly be of some management use. However, it is unlikely that these findings—or any management findings for that matter—will have much influence on future environmental policy in the Bay region.

The problem is not simply due to poor or unreliable data. The fact is that even if all the results were beyond reproach, they would be unlikely to have much policy influence. As long as the accounting exercise is viewed as an experimental research effort, all findings are likely to be similarly viewed as "academic." Only when the accounting activity becomes institutionalized within the bureaucratic structure will it gain influence over bureaucratic decisions. For this reason, the current effort toward implementing the Peskin framework in the Philippines involves several measures to help assure institutionalization. First, the project directly involves personnel from several governmental ministries. Second, besides the effort to build the accounts, the project is generating several prototypical policy studies in order to illustrate the usefulness of the accounting data bases. Finally, the project includes a training program in the techniques of environmental economics in order to enhance policy-making skills. This last step will both create a demand for the data bases and will improve their quality as the techniques of cost and benefit estimation get wider application.

Future Directions

This paper has argued that the policy benefits of "green" accounting will not be fully realized as long as the focus of the effort is solely on the generation of improved GDP and NDP income measures. More important than improved scorekeeping measures is the generation of a consistent integrated economic-environmental data base that efficiently supports the management of environmental and economic policy.

If, however, the management potential is to be fully achieved and resource and environmental accounting is to be more than of purely academic interest, the question of *where* "green" accounting should be done may be as important as *how* "green" accounting should be done.

The first concerted efforts at "green" accounting in the United States were essentially housed in non-governmental institutions such as universities and research institutes, e.g., the National Bureau of Economic Research (NBER), World Resources Institute, and Resources for the Future (RFF).[6] As long as the effort is perceived as experimental, maintaining some separation from "official" governmental statistical activities makes sense. Initial efforts necessarily require the publication of rough estimates, which could be open to misinterpretation were they published by an official agency.

Offsetting the research advantage of not being too close to government are two disadvantages. One is the fact that the researchers are often denied access to relevant but confidential government statistics. This denial is not only inconvenient but also expensive. For example, the NBER and RFF efforts would have been far easier if the researchers were not forced to develop on their own address lists of industrial establishments—lists that are available at the Department of Census but are considered confidential. A second disadvantage is that the "green" accounting effort will never attain the same visibility as an instrument of policy as it would be were it formally institutionalized within government.[7]

More recently, there appears to be more interest in the United States in doing "green" accounting within the federal bureaucracy. Besides the Chesapeake Bay regional accounting discussed in this paper, there are currently small efforts underway at the Bureau of Economic Analysis (BEA) and at the EPA. It is not clear which of these two agencies has an inherent advantage.

In BEA's favor, there is access to confidential census data. Moreover, as the agency with the official responsibility for national economic accounting, housing the effort within BEA would give the effort high visibility and respectability among mainstream economists. On the other hand, BEA has no special access to environmental data. Nor does it have a staff with the experience in the benefit-cost techniques needed to impute values for environmental services. EPA clearly has the advantage in these two areas. Moreover, because of its official responsibility as the national economic accountant, BEA would have a natural reluctance to publish any imputations, let alone the especially crude imputations that would be expected in the early years of "green" accounting. As many of these imputations would rely on benefit-cost analyses of existing and proposed environmental regulations, EPA would presumably be less reluctant to publish them.

The ultimate key to institutionalization is the use of the accounting system for policy making. While this fact argues for a leadership role for EPA, several other agencies were also "clients" of the "green" accounting efforts at the NBER and at RFF. Indeed, it could be argued that non-EPA agencies would find the data more credible and, therefore, would be more willing to use the data were it viewed as a product of an official statistical agency.

Given that there is no clear case for any agency to take the lead in "green" accounting, one can only hope that those agencies that have expressed interest will be willing to work together in a spirit of cooperation. While some competition between agencies may be worthwhile, lack of cooperation can increase data development costs. It would be a shame if apparent inefficiency and high cost led to the demise of "green" accounting well before there was any chance for its institutionalization.

References

Arrow, K. J. 1963. *Social Choice and Individual Values*. 2nd edition. Wiley, New York, NY.

Baumol, W. J. and W. Oates. 1988. *The Theory of Environmental Policy*. 2nd edition. Prentice-Hall, Inc., Englewood Cliffs, NY.

Bockstael, N. E., K. E. McConnell, and I.E. Strand. 1988. *Benefits from Improvements in Chesapeake Bay Water Quality, Vol. 2*. Office of Policy and Resource Management report under contract Cr-811043-01-0. U.S. Environmental Protection Agency, Washington, DC.

Daly, Herman E. and J. B. Cobb, Jr. 1989. *For the Common Good: Redirecting the Economy Toward Community. The Environment, and a Sustainable Future*. Beacon Press, Boston, MA.

Dennison, E. F. 1979. Pollution abatement programs: Estimates of their effect upon output per unit of input, 1975–78. *Survey of Current Business* 59(8 pt. I):58–59.

ENRAP 1994. *Philippine Environmental and Natural Resources Accounting Project, Phase II*. Main report to U.S. Agency for International Development under contract 492-0432-C-00-1015-05, Manila, Philippines.

Grambsch, A. and R. G. Michaels with H. M. Peskin. 1993. Taking stock of nature: Environmental accounting for Chesapeake Bay. In Lutz, E. (ed.), *Toward Improved Accounting for the Environment*. The World Bank, Washington, DC.

Grambsch, A. and R. G. Michaels. 1994. The United Nations Integrated Environmental and Economic Accounting System: Is It Right for the U.S.? Paper presented to the 1994 AERE Workshop "Integrating the Environment and the Economy: Sustainable Development and Economic/Ecological Modelling," June 6.

Hamilton, K. 1994. *Pollution and Pollution Abatement in the National Accounts*. Unpublished draft, The World Bank, Environment Department (October).

Hicks, J. R. 1940. The valuation of the social income. *Economica* VII(2):105–124.

Japan, Economic Council, NNW Measurement Committee. 1973. *Measuring Net National Welfare of Japan*.

Jorgenson, D.W. and P.J. Wilcoxen. 1990. Environmental regulation and U.S. economic growth. *Rand Journal of Economics* 21(2).

Nordhaus, W. and J. Tobin. 1973. Is growth obsolete? In M. Moss (ed.), *The Measurement of Economic and Social Performance, Studies in Income and Wealth* vol. 38. Columbia University Press for the National Bureau of Economic Research. NY.

Peskin, H. M. 1991. Alternative environmental and resource accounting approaches. In Robert Costanza (ed.). *Ecological Economics: The Science and Management of Sustainability.* Columbia University Press. New York. NY.

Repetto. R.. W. Magrath. M. Wells. C. Beer. and F. Rossini. 1989. *Wasting Assets: Natural Resources in the National Income Accounts.* World Resources Institute. Washington. DC.

United Nations. 1993. *Handbook of National Accounting: Integrated Environmental and Economic Accounting.* Series F. No. 61. U.N. Department for Economic and Social Information and Policy Analysis. Statistical Division. New York. NY.

Notes

1. The difference between GNP and GDP concerns the location and ownership of production. GNP covers all goods and services produced by a country's nationals regardless of where the the production takes place—at home or abroad. GDP covers all production taking place within national borders regardless of the nationality of the producers—national or foreign. Either measure. in order to prevent double counting, excludes the production of goods that satisfy the current account input needs of producers themselves.

2. Specifically. depreciation estimates equaled the current year net reduction in the physical units of the asset times the (implicit) rental value of each unit. This calculation. which ignores both the interest rate and the lifetime of the asset. yields a correct calculation of economic depreciation only if unit rents increase exactly at the market rate of interest. For this assumption to be valid. all capital markets would have to be fully open and competitive such that all would yield the same return. Other more direct estimates of natural resource depreciation. which do not rely on assumptions of perfect capital markers. are orders of magnitude smaller. (See ENRAP 1994. Appendices D.E. and F.)

3. GDP was adjusted downward by the estimated cost to maintain the environment at a non-damaged level. i.e., a level of total pollution control. As long as the marginal social costs of reaching such a level of control exceeds the resulting marginal social benefits. attaining such a level would not be socially optimal.

4. Economic theory demonstrates that the *economic* value of a good or service—and. if marketed. its price—depends on the intrinsic

value of the last, or "marginal", unit supplied. If a good or service were in excess supply, the value of this marginal unit would be zero. Since supply is in excess, the marginal unit provides no additional benefit to the consumer.

5.   Even rarer are studies that analyze how Bay conditions affect human behavior. One exception is Bockstael, McConnell, and Strand 1988.

6.   The exception was the pollution abatement cost accounting activities of the Bureau of Economic Analysis, U.S. Department of Commerce.

7.   However, past efforts did make some, albeit less visible, policy contributions. The EPA, US Department of Agriculture, the NOAA, and the U.S. Geologic Survey did make use of the underlying data bases occasionally for which they provided much needed financial support for data development.

CHAPTER NINE

Institutional Design for the Management of Estuarine Ecosystems: The Chesapeake Bay[1]

Timothy M. Hennessey

The Challenge of Managing Estuarine Ecosystems

This paper analyzes the Chesapeake Bay Program, the governance system that manages efforts to protect and restore Chesapeake Bay, the largest and richest estuary in the United States.

Estuaries are coastal water bodies containing seawater diluted by freshwater from fluvial sources. They are among the most productive and complex types of ecosystems (Odum 1983). Besides serving as excellent wildlife habitats, estuarine regions are home to millions of people. Human interaction with the estuary and its associated watershed have had both direct and indirect impacts on these ecosystems. These impacts come from a wide array of human activities such as farming, fishing, commercial shipping, recreational activities, waste disposal, industrial pollution, and coastal development. The cumulative impacts of human activity require effective management mechanisms in order to minimize such impacts. If these estuarine ecosystems are to be considered as a part of the human environment and are to be managed in order to attain recognized objectives, then the dynamics of the ecosystem and the human uses of it must inform the design of these institutions and management systems (Dryzek 1987a, 1987b, 1987c; Goodall 1977). Moreover, the distinctive characteristics of each marine ecosystem present unique problems for the design and management of governance institutions (Dryzek 1987c).

The science of ecology informs our understanding of such ecosystems, including estuarine ecosystems. Ecologists investigate what happens to organisms, populations, and communities under particular sets of circumstances. At the level of organisms, ecology deals with how individuals are affected by or how they affect their environment. At the level of population, it deals with the presence or absence of particular species and with fluctuations and trends in their numbers. Community ecology deals with the composition and structure of communities and the pathways followed by nutrients, energy, and chemicals as they pass through the communities. These communities are not constant but are continually changing because of interactions among the populations and because of disturbances from climactic and geological events and human intervention (Odom 1983; Begon, Harper, and Townsend 1986).

Ecologists have identified three general characteristics of ecosystems that have implications for the management of estuaries. First, the elements of ecosystems are highly interdependent; that is, "everything is connected to everything else" (Commoner 1972). Moreover, there are few linear relationships in the system: "Every effect is also a cause in a web of mutual interdependency" (Ophuls 1977). These properties constrain our ability to determine the boundaries of particular ecosystems and to establish basic causal relationships. Second, ecosystems are nonreducible. They are dynamic, interconnected, systems which cannot be broken down into simpler parts (Edson, Foin, and Knapp 1981). Thus the "health" of an estuary cannot be predicted with certainty using information on selected parts, and the overall problems of an ecosystem cannot be solved through the improvement in the condition of its component parts. As Dryzek observes, "Non-reducibility in ecological problems can make the way we customarily allocate problems into separate issue areas look suspiciously like a bad case of tunnel vision." (Dryzek 1987c) Third, ecosystems are characterized by homeostasis, adaptiveness, and succession, which lead to great temporal and spacial variability. These properties guarantee that the context of the problem will not remain fixed.

These properties of ecosystems create problems of extreme uncertainty. The more the system is interconnected, non-reducible, and variable, the more difficult it is to isolate its component parts for management purposes. Moreover, the more dynamic the system, the more difficult it is to model its current state or predict its future course. In short, the more complex the system, the more difficult it is to understand and manage. Many of these analytic difficulties associated with ecosystems generally and estuarine ecosystems in particular are revealed in our analysis of the development and evolution of The Chesapeake Bay Program.

Responding to Complexity and Uncertainty: The Role of Adaptive Management

Given these characteristics of complexity and uncertainty, the management program design must incorporate the capacity to learn in order to adjust to new information as this becomes available. Adaptive management is the most promising approach to cope with such uncertainty. Adaptive management takes such uncertainty seriously by treating human interventions in ecosystems as experimental probes. As Lee points out, "Its practitioners take special care with information. First they are explicit about what they expect, so they can use methods and apparatus to make measurements. Second they collect and analyze information so that expectations can be compared with reality. Finally they transform comparison into learning—they correct errors, improve their imperfect understanding and change action and plans. Linking science and human purpose, adaptive management serves as a compass for us to use in searching for a sustainable future" (Lee 1993).

Adaptive management is learning by doing—by treating programs as experiments. This approach yields what Lee calls "reliable knowledge." In this sense, the implementation of the program creates opportunities to test and improve the scientific basis for action. Adaptive management is built on five working assumptions: "1. The purpose of adaptive management is the protection and restoration of living resources. 2. Projects are experiments: the choice is to make them good ones or bad ones. 3. Action is overdue. We do not delay action until enough is known. 4. Information has value, not only as a basis for action but also as product of action. 5. Protection measures may be limited but management is forever." (Lee and Lawrence 1986 and, in general, Holling 1978; Walters 1986). In short, an adaptive policy is one that is designed from the start to test hypotheses about the behavior of the ecosystem impacted by human use.

But adaptive management is not enough to assure that large-scale ecosystems are managed sustainably. Lee argues persuasively that the other necessary element is bounded conflict (Lee 1993). The latter is fundamental to the management of such systems because what makes ecosystems difficult to manage is not just geographic size per se but interdependence of use across several jurisdictions with large populations. In this sense, "large ecosystems are arenas of interdependence which inevitably produce conflict which, in turn, produce opportunities for institutional invention and learning." (Lee 1993: 11)

Much of this conflict revolves around the demand and use of collective goods. Common property goods such as fisheries present a problem because, in the absence of an effective management regime, rational individuals will tend to overharvest the resource and thereby produce a negative result for society referred to by Hardin (1968) and others as the "Tragedy of the Commons." Problems of a similar nature include the assimilative capacity of rivers and estuaries, groundwater basins, and the protection of natural environments (Dryzek 1987c). Lee's point is that such conflict can provide ways to recognize errors, thereby complementing and reinforcing the self-conscious learning of adaptive management (Lee 1993). For Lee, policy learning is a by-product of competition among political actors (i.e. experts, politicians and bureaucrats). In what follows, I shall use adaptive management and bounded conflict to analyze the evolution of The Chesapeake Bay Program.

Selecting a Governance System for Chesapeake Bay

Governance includes not only the laws, regulations, and programs for environmental control and the various uses of the estuary, but also the key actors and organizations that determine and implement such laws, regulations, and programs (Sproule-Jones 1980; Bish 1982). The nature of the ecosystem and the incentive system associated with human usage of it present a dual challenge to the design and management of the ecosystems. The governance system must not only manage individual human uses of the ecosystem, but must account for the interrelationships that exist between human uses and the impact of these on the ecosystem. Because of this interdependence of uses and users, estuary management must consist of more than passing data from "scientists" to "decision makers." Management occurs within an institutional setting that more or less successfully serves to reconcile the differing values and objectives of a variety of user groups and the general public and then provides the means for implementing chosen objectives. I refer to this institutional framework and process as the governance system.

In what follows, I trace the development of the Chesapeake Bay governance system. In 1975, Congress authorized a 5-year, $25-million study of Chesapeake Bay. The study was undertaken to estimate the nature and magnitude of threats to the Bay ecosystem. It had five basic objectives: (a) to assess the principal factors having an adverse impact on environmental quality; (b) to establish mechanisms for collecting, storing, analyzing, and disseminating environmental data; (c) to analyze available environmental data and implement methods for improved data collection; (d) to propose alternative control strategies for long-term protection of the bay; and (e) to evaluate bay management coordinating mechanisms.

In 1976, the Environmental Protection Agency (EPA) acted in accordance with congressional directives, establishing the Chesapeake Bay Program Office and an organizational structure to manage related activities. In 1981, a research team of Bay scientists was established to synthesize data from the 40 individual scientific studies. This led to a summary report of technical studies in 1983 that synthesized the scientific understanding of the Bay and concluded that the Chesapeake was in trouble. The striped bass and shad fisheries had declined sharply during the period. Bay sea grasses, a critical component of wildlife habitat, had become scarce. Nutrients from phosphorus and nitrogen were overloading the bay waters, creating algal blooms that diminished oxygen supply to the ecosystem. Toxics also were found at alarming levels in bottom sediments (USEPA 1983a, 1983b).

In addition to scientific inquiry into the condition of the estuary, Congress directed EPA to determine which units of government should have management responsibility for the environmental quality of the Bay and to define how such management responsibility should be structured so that communications and coordination could be improved among units of government, among government units and research and educational institutions, and among government units and concerned groups and individuals on Chesapeake Bay. Because the Bay is a complex, interactive system, actions taken in any part of the watershed may affect a downstream environment. It was therefore critical to ensure a cooperative effort among the governing agencies. This fundamental approach dominated deliberations concerning the appropriate governance system for Chesapeake Bay circa 1979. From 1979 until 1983, a number of options for alternative institutional arrangements were given careful consideration. Two major consulting reports formed the basis for informed choice making (Capper et al. 1981; Resources for the Future [RFF] 1979).

Capper et al. identifies the institutions involved in Chesapeake Bay evaluation and management, tracing the early efforts to manage the Bay. They note that in 1978, the general assemblies of Maryland and Virginia created the Chesapeake Bay Legislative Advisory Commission to evaluate existing and potential management institutions for the Bay. The commission reviewed several general types of alternative management institutions.

They conclude that primary responsibility for governing Chesapeake Bay should remain with the states and their political subdivisions. Management difficulties arising from intra- and interstate jurisdictional boundaries should be resolved through cooperative efforts. Accordingly, in 1980, the Maryland and Virginia General Assemblies created the Chesapeake Bay Commission, which consisted primarily of legislative members from both states, with one executive agency and one citizen member from each state.

After the Capper Report, fundamental questions remained: namely, was a new institution for the Bay needed, and if so, were any of the existing institutions noted above suitable for carrying out the specified activities and responsibilities? And what other alternatives existed to answer these questions? To address these problems, the Chesapeake Bay Program commissioned Resources for the Future, Inc. (RFF) to evaluate institutional arrangements for water resource problems (RFF 1979). Their report examined alternative water management arrangements used to accomplish regional environmental objectives.

The report concluded that regional institutions in the United States did not perform as expected in solving the problems that they were designed to address, primarily because existing local, state, and federal entities tended to resist new regional institutions. Their exercise of authority tended to be limited, regardless of how strong their authority appears on paper.

The RFF report did not select the "best" institutional arrangement, but identified public choice criteria to be used in designing new regional institutions for estuarine ecosystem management. Their arguments reflect research findings in economics, political science, game theory, and law relating to the criteria for designing management institutions for natural resources (Kiser and Ostrom 1982; Oakerson 1986; Ostrom 1990; Tang 1992). Among the key factors this research identified are the physical attributes of the system (i.e., the estuarine ecosystem), the attributes of the community of users, and the type of institutional arrangements chosen. These factors combine to create different incentives and constraints for users. The users, operating under conditions of uncertainty, behave strategically as they react to the opportunities inherent in the situation, thereby producing different outcomes. The long-term viability and performance of the management system depends on the institution's ability to acquire and process the information necessary for effective operation and maintenance of the ecosystem. High levels of uncertainty presented a particular challenge for institutional designers in the Chesapeake.

Responding to this challenge, RFF analysts argued that the jurisdictional scope of the institutions should correspond or "fit" with the impact boundaries of problems insofar as adequate knowledge about impact boundaries is available. That is to say that the institutions selected should be at the scale of the problem and not too large or too small in scope. Moreover, following Williamson (1975), they argued that small institutions are more efficient and responsive than large institutions because they respond to citizen preferences more readily. Therefore, institutions in the Chesapeake should be no larger than the scale required to incorporate the preferences of all relevant parties affected by the problem.

RFF noted that a multiple-institutional governance system is preferred for dealing with problems having potentially serious consequences where adequate information about the impact boundaries of those problems is not available. RFF observed that where very little is known about the effects of a problem, the most important function an institution can perform is to collect and generate new information about the problem, and that unbiased information can best be collected and generated through a multiple-institution structure.

They concluded that the cost of a multiple-institution structure for collecting and generating new information about a problem is justified if the problem has potentially serious consequences; that is, complex problems are best handled by a number of institutions rather than a single institution which would have fewer sources of information, a low capacity for adjustment, and therefore would be more error prone. The benefits of a number of different perspectives on complex problems outweigh the potential inefficiencies of a multiple-institution structure.

Finally RFF argued that creating a new institution is feasible only if a favorable incentive structure exists for the participants in that institution: the prospective members must believe that the advantages outweigh the disadvantages of participating in the institution.

The RFF study suggested that management institutions should correspond to the states and municipalities that drain or border the Bay. Because the Bay drainage area encompasses significant portions of five states and the District of Columbia, the institution should encompass either all five states and the District of Columbia or at least those states whose land area or volume of effluent can significantly affect the Bay and its tributaries. The report recommended that existing mechanisms should be tried until it is demonstrated that those mechanisms cannot achieve the desired objectives (USEPA 1983a).

Using the Capper et al. and RFF reports as a point of departure, the Chesapeake Bay Program considered three classes of action. The first, that of using existing structures, contained two options (1 and 2 below); the second, to modify an existing institution, contained six options (3–8 below); the last, to create a new institution, contained two options (9 and 10 below). The ten options are as follows: (1) EPA Region III; (2) EPA Region II and the Chesapeake Bay Program Management Committee; (3) Chesapeake Bay Policy Board and Management Committee; (4) Bi-State Working Committee; (5) Chesapeake Bay Commission; (6) Interstate Commission on Potomac River Basin; (7) Susquehanna River Basin Commission; (8) Chesapeake Research Coordination Board; (9) Basin Commission; (10) Comprehensive Bay-Wide Authority.

To make maximum use of current programs, the option of EPA or EPA and the Chesapeake Bay Committee looked promising. And EPA did indeed have a great deal of authority to negotiate tough state and EPA agreements to revise water quality standards and force funding priorities for nutrient loadings and the like. But some of the CBP recommendations fell outside EPA's jurisdiction, especially in the areas of storm water management and fisheries. EPA had no authority to implement programs in these two critical areas. And although EPA had formal authority under the Clean Water Act to enforce abatement programs in the region, formalized enforcement activities could be long and tedious because of checks and balances between the federal government and the states.

The combination of the Chesapeake Bay Committee (CBC) with EPA also looked like a promising combination because the CBC had been very effective in providing program guidance to the EPA on an informal level, which resulted in an effective approach to management of a complex federal/state program. The management committee also oversaw the implementation of a Chesapeake Bay Data Management Center and encouraged the initiation of monitoring activities recommended by EPA. The Management Committee in the CBC could be modified to include representatives of state and federal agencies with responsibilities for managing Bay resources. Hence, the combination of the EPA plus the CBC looked like it might be a viable option.

The second category contained six options of institutions which already existed and might be modified to carry out the required tasks. Analysts agreed that any such institution, if modified, should be able to coordinate water quality and resource management programs at the state and federal level; be structured to include federal agencies such as EPA, NOAA, USDA, and others, the states of Virginia, Maryland, Pennsylvania, and the District of Columbia, as well as local governments; and have an advanced technical capability, be able to handle oversight of monitoring programs, maintain and use large scale computer models, and carry out public outreach (USEPA 1981).

Part of any such institution should be the greater use of citizens and the technical advisory committees in order to meet the criteria above. A Chesapeake Bay Policy Board and Management Committee was suggested as one mechanism to achieve this. The Policy Board would have representatives appointed by the governors of the three states and the District of Columbia. The chair would be the regional administrator of EPA. It would maintain a strong role for the federal government and the affected states, set policy and make resource allocation decisions at the highest levels, accommodate the diverse needs of the federal, state and local governments, and provide a continuing forum for discussion and resolution of issues and disputes affecting the ecological health of the Bay.

The Management Committee would be charged with implementing these strategies, and would be composed of two representatives each from the states and the district with water quality and resource management responsibilities, and one each from the federal government agencies with regulatory responsibilities in the Bay. The Chairman would be the director of the Region III Water Division. It would carry out plans, implement monitoring programs, develop regional approaches to water quality standards and waste load allocations, and establish priorities for funding. The committee would also review ongoing Bay research and recommend research on specific issues. As we shall see, a version of The Chesapeake Bay Policy Board and Management Committee eventually became an essential element of the Chesapeake Bay Program.

Several institutions already in existence were considered. The Bi-State Working Committee was an option, but this committee from Maryland and Virginia was not adequate to handle issues for the whole Bay because it would need to add the federal government and Pennsylvania and District of Columbia members. Moreover, it had no full-time staff or scientific capability. The Chesapeake Bay Commission was also considered but presented problems similar to the Bi-State Committee. It was purely advisory to Maryland and Virginia with respect to achieving legislative solutions to mutual problems. It had no capability to carry out programs or scientific staff. Moreover, it would have to change its charter to include Pennsylvania, the District of Columbia, and the Federal government.

The Interstate Commission on The Potomac River was also considered, but it dealt only with problems of the Potomac River Basin and would, therefore, require fundamental changes to deal with the Bay as a whole. While it already had the necessary membership, it, like the others, was without operational or scientific capability, and modification of its compact would also have required congressional action, taking several years. The Chesapeake Research Coordination Board was also considered, but it dealt solely with research coordination and was not equipped to handle the needs of strategic planning and oversight of monitoring. It would also need to add Pennsylvania to the board, which would require an act of Congress. The final option in this category, The Susquehanna River Basin Commission, recognized the river's influence on the Bay and did have a regulatory, management, planning and coordination function and a limited technical capacity. The SRBC participants included Maryland, Pennsylvania and New York. The state of Virginia and the District of Columbia would have had to be added to obtain a Bay-wide focus. Moreover, changes in its compact would have to be approved by each state and the federal government which could have taken several years (USEPA 1983a).

The last set of options considered were new institutions to be set up for the express purpose of Bay-wide management. As early as 1977, Senator Charles Mathius suggested that a single Bay-wide institution be created. He suggested a Basin Commission created under Title II of the Water Resources Planning Act of 1965. He felt that a single agency was needed to coordinate all the federal, state, local, and private agencies with interests in the Bay. He estimated that there were ten federal agencies with jurisdiction in the Bay, five interstate agencies and commissions, 31 state agencies in Maryland and Virginia dealing with the Bay, and seven Maryland and Virginia colleges and universities studying the Bay. Advantages of such a commission were that it would have required state and federal participation, have the governors with states affecting the Bay involved, have interstate authority, have authority to carry out needed programs, and have the capacity to recommend necessary research.

The other option was the creation of a comprehensive Bay-wide authority to manage the ecosystem and its resources. The authority would have all of the necessary powers and resources to manage and coordinate all aspects of the Bay as a national resource. This would, however, require congressional action to create a new federal agency with broad responsibilities.

The Chair of the Chesapeake Bay Program Management Committee and the director of the Chesapeake Bay Program concluded that the need for immediate action and the costs involved made it essential that an existing mechanism with basin-wide federal state representation be responsible for coordinating the cleanup (USEPA 1983a). They recommended Option 2, EPA Region III and the Management Committee, as the preferred option.

They also recommended that the Management Committee be the coordinating mechanism. The committee periodically was to brief the EPA regional administrator and state secretaries or their equivalent. Specific responsibilities of the committee included coordinating the implementation of the Chesapeake Bay Program recommendations; developing a comprehensive basin-wide planning process in conjunction with ongoing planning efforts; investigating new regional approaches to water quality management, including creative financing mechanisms; resolving regional conflicts regarding water quality issues; and reviewing ongoing Bay research efforts and recommending additional research needs (USEPA 1983a).

The Chesapeake Bay office was to maintain the computer base, coordinate the monitoring program, and refine the Bay water quality models. Public participation and effective coordination among and between levels of government was to be undertaken as part of the management efforts.

The 1983 Chesapeake Bay Agreement

These recommendations formed the basis for the governance structure of the 1983 Chesapeake Bay agreement signed by the EPA and the states of Virginia, Maryland, Pennsylvania, and the District of Columbia. The agreement established the major elements of a cooperative structure to develop and coordinate the comprehensive bay cleanup: namely, the Chesapeake Bay office.

The council membership included representatives from each of the four jurisdictions and from the EPA. Chairmanship of the council went to one of the three state governors, the mayor of the District of Columbia, or the representative of the EPA. Operating by consensus, the council's primary functions were planning and coordination to ensure efficient implementation of programs and projects to restore the Bay.

The Implementation Committee, the council's operating arm, had 26 members: delegates from the jurisdictions and representatives of the seven federal agencies and three interstate commissions (Chesapeake Bay Commission, Interstate Commission on the Potomac River Basin, and Susquehanna River Basin Commission). Subcommittees for Planning, Nonpoint Sources, Data Management, Modeling and Research, Monitoring, and Living Resources were to coordinate work in those categories across agency and state lines. A Scientific and Technical Advisory Committee (STAC), whose membership included directors of major Bay area research institutions, also assisted the Implementation Committee. The Chesapeake Research Consortium, an organization of bay research institutions, provided support for STAC through an EPA grant.

The council had a Citizens Advisory Committee (CAC) to provide a public perspective on policy issues. CAC had 25 members: four appointed by the chief executive in each state and nine at-large members nominated by the Citizens Program for the Chesapeake Bay, Inc. (Figure 9–1).

The Chesapeake Bay Restoration and Protection Plan of 1985

In 1985, the four jurisdictions and seven federal agencies examined the programs and produced a catalogue of goals for the Bay system based on recommendations from the Resource Users Management Team (USEPA 1983c). This Chesapeake Bay Restoration and Protection Plan (1985) was the first clear statement of specific goals and a linkage of these goals to state programs. The overarching policy was: "to improve and protect the water quality and living resources of the Chesapeake Bay estuarine system so as to restore and maintain the Bay's ecological integrity, productivity and beneficial uses and to protect public health" (Chesapeake Executive Council 1985a).

Five specific goals were identified and tied to state and federal program activities, as depicted in Figure 9–2. The general goals were (a) *nutrients:* to reduce point and nonpoint nutrient loadings to attain nutrient and dissolved oxygen concentrations necessary to support the living resources of the Bay; (b) *toxic substances control:* to reduce or control point and nonpoint sources of toxic materials to attain or maintain levels of toxicants not harmful to humans or living resources of the bay; (c) *living resources:* to provide for the restoration and protection of the living resources, their habitats, and ecological relationships; and (d) and (e) *institutional and management activities:* to support and enhance a cooperative approach toward Bay management at all levels of government.

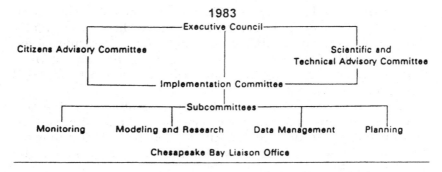

Figure 9–1. Chesapeake Bay Program management structure (as of 1983).

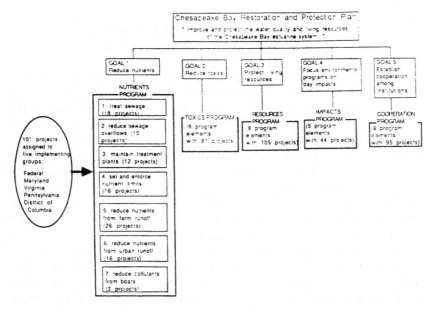

Figure 9–2. The implementation structure for the Chesapeake Bay Program (as of 1985).

The multiplicity of programs and projects that are required to implement the policies and the number of decision makers associated with each of these give a sense of the complexity involved in the implementation process.

As Figure 9–2 indicates, the overall policy had five goals, each of which drew on a number of programs—32 in all. These programs, in turn, were subdivided into 430 projects to be administered by five jurisdictions: Maryland, Virginia, Pennsylvania, the District of Columbia, and federal agencies. As an illustration of the complexity involved in Figure 9–2, the policy goal of nutrients has been broken down into its program elements and these, in turn, into projects, although space restrictions prevent us from listing the 101 projects needed to implement this one particular goals across five governmental jurisdictions!

The 1987 Agreement: Evolution of Goals and the Expansion of the Governance Structure

A continuing process of learning, adaptation, and evolution in the program is clearly demonstrated in the updating of the Chesapeake Bay Agreement in 1987. In January 1987, Virginia Governor Gerald Baliles, Chair of the Chesapeake Executive Council, proposed that the adequacy of the 1983 agreement be examined and that a new agreement be developed if necessary. In August 1987, a draft agreement was released and a public review process launched (Chesapeake Executive Council 1987b). That process, along with the information developed for the report, helped the drafting committee to revise and complete the new Bay agreement which was signed on December 15, 1987.

Based on a series of reports and much discussion with stakeholders and among themselves, the Programs Implementation Committee, the 30-member group that meets regularly through the year as the operational arm of the Executive Council, established new subcommittees for Toxics, Population Growth and Development, Public Access, and Public Information and Education. The Committee retained four other subcommittees—Living Resources, Monitoring, Modeling, and Nonpoint Source—and the Federal Agencies Committee, but eliminated a Data Management Subcommittee and divided its responsibilities between the Monitoring and Living Resources panels. Other elements of the Bay program included a Budget Committee, Workplan Steering Committee, and a 1991 Nutrient Reevaluation Workgroup to carry out the reassessment of the 40 percent nutrient reduction goal called for by the agreement. In addition, the Citizens Advisory Committee, the Local Government Advisory Committee, and the Scientific and Technical Committee brought the perspectives of their constituencies to the committees and the workgroups of the Bay program.

Several of the features of the agreement should be highlighted because they demonstrate the learning capacity of the governance system. For example, the issues of public access and growth were added since the 1983 agreement in recognition of the need to promote opportunities for public access to the Bay and the need to mitigate the adverse effects of continued growth and development.

The government advisory committee was added to acknowledge the key role played by local government in the management of nonpoint and point sources of pollution (see Figure 9–3). Clearly, these activities cannot be successfully implemented without local support (Barker 1990; Favert et al. 1988; Warner and Warren 1985).

The agreement also recognized the critical importance of the nutrient strategy and the specific goal of a 40 percent reduction by the year 2000 in nitrogen and phosphorus entering the mainstream of the Bay. The nutrient reduction strategy was arrived at between 1983 and 1987 based on a mathematical model which established targets for reducing the amount of nitrogen and

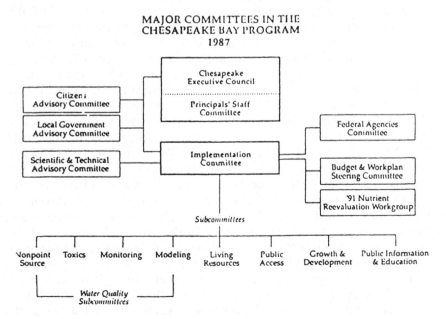

Figure 9–3. Major committees in the Chesapeake Bay Program (as of 1987).

phosphorous entering the Bay. The model evaluated the water quality response of the Bay to a variety of nutrient reduction scenarios. The model's results predicted that if nutrient loads were reduced 40 percent, nutrient enrichment would be reduced sufficiently to stop the depletion of dissolved oxygen, thereby encouraging the recovery of the Bay's living resources to higher population levels. The nutrient issue was considered so important that it was included as a work group reporting directly to the Implementation Committee and charged with a full-scale reevaluation of the nutrient goal by 1991.

This new agreement, in contrast to the 1983 agreement, contains eight goals, 40 objectives, and 29 priority commitments with deadlines for managing programs in living resources, water quality, public information education and participation, population growth and development, public access, and governance. Many of the 29 commitments specified in the 1987 agreement were completed during the period from 1988 to 1992 (Hennessey 1994).

Science and Governance Program Implementation and Evaluation

After 1987, concern grew among the participants in the management process to know *how much more needed to be done, where, at what cost, with what results, and how much longer.* To help answer these questions, the Chesapeake Executive Council adopted a program evaluation and development process (Chesapeake Implementation Committee 1988). It was at this point in the management process that the decision makers seriously confronted the irreducibility and complexity problem inherent in ecosystems which was discussed in section one of this paper—that is, the problem of inferring the condition of the whole ecosystem with information on selected parts of the system. Decision makers in the Bay program responded to this challenge by designing a program for indicator species as a way to cope with the complexity. Instead of trying to monitor all the dimensions of the ecological whole, they focused on specific living resources and their habitat that are sensitive to an important cross-section of those dimensions. What the decision makers seemed to understand is that "indicator species are messengers of the well-being or ill health of the whole ecosystem" (Lee 1993). The Executive Council reasoned that to provide for the restoration and protection of living resources, their habitats, and their ecological relationships, it was necessary to set regional habitat goals; that is, the water quality, biological, and physical requirements necessary for continued propagation of representative living resources within a defined geographic area. The overall importance of this approach to evaluating the Bay program was acknowledged: "These regional habitat objectives were designed to guide overall management of the Bay and provide useful measures of restoration progress. *The ultimate measure of success will be the responses of living resources throughout the Bay*" (italics mine) (Chesapeake Implementation Committee 1988). Because living resources are the most tangible sign of widespread environmental problems, they were selected to serve as powerful analytic surrogates for building a chain of evaluative inferences about progress toward restoring the Chesapeake Bay ecosystem.

But as of 1991, the crucial linkages between living resources and other ecosystem factors had not been established. A report to Congress cited the need to better understand and quantify relationships and interactions between components of the ecosystem so that the bay can be managed as an integrated ecosystem. More specifically, the report focused on (a) the relationship between the lower and higher forms of living resources, (b) whether meaningful links could be made with the Bay Water Quality Model between water quality and higher forms of living resources, and (c) the interplay between living resources and water quality (Chesapeake Bay Program 1991).

Although the precise linkages between water quality and living resources are still unclear, this should not diminish the record of progress in particular programs in the Chesapeake Bay Program as of August 1991. For example, there is a continuing decrease in the levels of phosphorus entering the Bay. Total levels of phosphorus are down 20 percent since 1985, and dissolved organic phosphorus is down 41 percent since 1985. These reductions allow a greater supply of oxygen to support plant and animal life. As of 1989, submerged aquatic vegetation was up to 57 percent from 1984 in the mid-Bay region, thereby providing improved habitat for living resources. Blue crabs appear relatively healthy, and striped bass (or rockfish) are increasing, again allowing sport and commercial fishing for these popular fish. This stock improvement permitted the lifting in 1990 of the 5-year fishing moratorium (Chesapeake Executive Council 1991). Important challenges still exist: dissolved oxygen has remained essentially unchanged from 1984 to 1990, and nitrogen in the Bay is still rising.

Summary, Conclusions, and Observations

The Chesapeake Bay Program is an adaptive system which addressed increasingly complex issues while integrating existing management mechanisms. These characteristics are evidenced in three stages of development: (a) 1976–1983—Agenda Setting: Science and Public Choice; (b) 1983–1986—Choice of Governance Structure and Management Initiatives; (c) 1987–1992—Science and Governance: Program Implementation and Evaluation.

In Stage 1 (1976–1983), the Resources Users Management Team served as the key public choice mechanism that brought together stakeholders and scientists to select the key issues facing the Bay in light of extensive scientific research.

In Stage 2 (1983–1986), the institutional structure of the program was established. In light of two consulting reports, a decision was made not to create a new regional authority to govern and manage the Bay. Instead, a multistate, federal cooperative program was selected owing to the fact that regional authorities tend to be resisted by existing local authorities. A multi-institutional design has jurisdictions that more closely correspond to problem parameters and provide more unbiased information than regional authorities. The agreement established the major elements of a cooperative structure to develop and coordinate the comprehensive Bay cleanup. In 1985, the four jurisdictions and the federal government produced the Chesapeake Bay Restoration and Protection Plan, the first clear statement of specific goals and linkage of these goals to state programs.

In Stage 3 (1987–1992), there was an expansion of goals and governance structure as a result of the 1987 Chesapeake Bay Agreement. This new agreement established eight goals, 40 objectives, and 29 specific commitments. The issues of public access and growth were added, the governance structure was expanded to include a local government advisory committee, and federal government agencies became full partners to the 1987 agreement. Finally, the new agreement recognized the critical importance of the nutrients by setting a goal of 40 percent reduction in nutrients by the year 2000. After 1987, the need for evaluation became apparent and living resources were chosen as the best indicator of Bay health. An evaluation program was constructed around regional habitat indicators in terms of which to build a chain of evaluative inferences to judge the state of the whole estuary.

The chain of evaluative inferences about progress toward restoration and protection of the Bay would be aided significantly if the Bay Program took more seriously the experimental aspects of adaptive management. The Chesapeake Program could do more to treat projects as experiments—it should have the power of testing. The establishment of the living resources/habitat program was an important step in that direction but was arrived at quite late in the process.

In short, the Chesapeake Bay Program has demonstrated a strong learning capacity despite the absence of experimental evidence, but this could be considerably enhanced by treating programs as rigorously designed experiments. If this were to be done, the excellent efforts toward adaptive management could be enhanced so as to assure a strong body of knowledge of the bay ecosystem to future managers.

Perhaps the most important lesson that can be learned from the Chesapeake Program is that multi-jurisdictional, cooperative management programs are successful institutional designs for managing complex estuarine ecosystems. These systems create incentives for local, state, and federal agencies to provide scientific information and to implement agreements that were arrived at in a cooperative manner. The institutions "bound" conflict by channeling it into information and preferences vital to our understanding of human use of ecosystems, and citizens at a variety of levels tend to participate actively and thereby develop an efficacious view of their role in the program. Such "ownership" is essential for the long run sustainability of such management institutions.

Finally, it is important to see the Chesapeake Program as a part of "a diffusion of innovations." Lessons learned in the Great Lakes Program were incorporated into the Chesapeake Program and lessons from the Chesapeake Bay Program have been incorporated in the National Estuary Program. A clear lesson that emerges from the experience of the Great Lakes Program was that pollution controls evolve in an adaptive phased pattern. In other words, pollution control programs build on the successes of past governance attempts (international agreements in the Great Lakes) while trying to correct the weaknesses and avoid the failures of past regulatory programs. The process involves identifying pollution problems based on impaired uses; linking those impairments to pollutants; isolating the origin of the primary pollutants; developing corrective actions; and implementing those corrective actions. In order to employ this type of approach, it became clear that it was necessary to involve a broad range of scientists, political actors, and concerned members of the public in order to assimilate and incorporate this information into management actions (Imperial, Robadue, and Hennessey 1993; Imperial, Robadue, Hennessey 1992).

The management lessons of the Chesapeake Bay Program in turn have been incorporated into the National Estuary Program, established by Congress in 1987 and thereby expanding EPA's use of regional management strategies to protect coastal environments. Starting in 1987 and up until 1993, 23 U.S. estuaries were in various stages of preparing a comprehensive conservation and management plan.

As in the Chesapeake Bay Program, the management strategy of the National Estuary Program incorporates federal, state, and local governments, the scientific community, and concerned members of the general public in an interactive and collaborative decision-making system called the Management Conference. The Management Conference stimulates the transfer of scientific, technical and management experience and knowledge among the Estuary Program participants. It also helps to enhance the general public's and the decision-maker's awareness of the environmental problems affecting the estuarine ecosystem and serves as the means to discuss and propose solutions to these problems. Thus, the Management Conference is intended to be both interactive and collaborative, and is focused on building partnerships among the estuary program participants. These partnerships are important in order to obtain the necessary commitments that are required for the effective implementation of a Comprehensive Conservation and Management Plan (CCMP).

In sum, the NEP is a culmination of management lessons learned first in the Great Lakes Program, which were refined in the Chesapeake Bay Program and are now incorporated into the management system of the National Estuary Program.

References

Barker, P. D., Jr. 1990. The Chesapeake Bay Preservation Act: The problem with state land regulation of interstate resources. *William and Mary Law Review* 31:735–72.

Begon, M., J. L. Harper and C. Townsend. 1986. *Ecology.* Sinauer, Sunderland, MA.

Bish, R. 1982. *Governing Puget Sound.* University of Washington Press, Seattle, WA.

Capper, J., G. Power, and F. Shivers, Jr. 1981. *Governing Chesapeake Waters: A History of Water Quality Controls of Chesapeake Bay, 1607–1972.* U.S. Environmental Protection Agency, Washington, DC.

Chesapeake Bay Living Resources Task Force. 1987. *Habitat Requirements for Chesapeake Bay Living Resources.* Annapolis, MD.

Chesapeake Bay Monitoring Subcommittee. Data Analysis Workgroup. 1989. *The State of Chesapeake Bay: Third Biennial Monitoring Report.* Annapolis, MD.

Chesapeake Bay Program. 1991. *Status and Workplan Report to Congress.* Washington, DC.

Chesapeake Executive Council. 1985a. Scientific and Technical Advisory Committee. *Chesapeake Bay Restoration and Protection Plan.* Annapolis, MD.

———. 1985b. *First Annual Progress Report Under the Chesapeake Bay Agreement.* Annapolis, MD.

———. 1987a. *Available Technology for the Control of Nutrient Pollution in the Chesapeake Bay Watershed.* Annapolis, MD.

———. 1987b. *Chesapeake Bay Agreement.* Draft.

———. 1987c. *Second Annual Progress Report Under the Chesapeake Bay Agreement.* Annapolis, MD.

———. 1988. *Population Growth and Development in the Chesapeake Bay Watershed to the Year 2020.* Annapolis, MD.

———. 1989. *The First Progress Report Under the 1987 Chesapeake Bay Agreement.* Annapolis, MD.

———. 1989. *The Second Progress Report Under the 1987 Chesapeake Bay Agreement.* Annapolis, MD.

———. 1991. *The Chesapeake Bay: A Progress Report 1990–91.* Annapolis, MD.

Chesapeake Implementation Committee. 1987. *Second Annual Monitoring Report: The State of the Chesapeake Bay.* Annapolis, MD.

Chesapeake Implementation Committee. 1988. *The Chesapeake Bay Program: A Commitment Renewed: Restoration Progress and the Course Ahead Under the 1987 Agreement.* Annapolis, MD.

Commoner, B. 1972. *The Closing Circle,* Knopf, New York, NY.

Dryzek, J. S. 1982. Policy analysis as a hermeneutic activity. *Policy Sciences* 14:309–329.

———. 1987a. Complexity and rationality in political life. *Political Studies* 35:424–442.

———. 1987b. Discursive designs: Critical theory and political institutions. *American Journal of Political Science* 31:656–679.

———. 1987c. *Rational Ecology.* Basil Blackwell, Oxford, UK.

Edson, M. M., T. Foin and C. Knapp. 1981. Emergent Properties and ecological research. *American Naturalist* 118:593–6.

Favert, P., D. Pitt and D. F. Tuthill. 1988. Land use policies, water quality, and the Chesapeake Bay. In G. Johnson and Philip Favert (eds.), *Natural Resources and Environmental Policy Analysis,* pp. 120–145. Westview Press, Boulder, CO.

Flemer, D. A., G. B. MacKernan, W. Nehlsen, V. K. Tippie, R. B. Biggs, D. Blaylock, N. H. Burger, L. C. Davidson, B. Haberman, K. S. Price, and J. L. Taft. 1983. *Chesapeake Bay: A Profile of Environmental Change.* U.S. Environmental Protection Agency, Chesapeake Bay Program, Washington, DC.

Goad, M. 1988. Save the bay. *Government Executive* February, 34–40.

Goodall, D.W. 1977. Dynamic changes in ecosystems and their study: The roles of induction and deduction. *Journal of Environmental Management* 5:309–317.

Hardin, G. 1968. The tragedy of the commons. *Science* 162:1243–8.

Healey, M. and T. Hennessey. 1994. The utilization of scientific information in the management of estuarine ecosystems. *Ocean and Coastal Management* 23:167–191.

Hennessey, T. M. 1994. Governance and adaptive management for estuarine ecosystems: The case of Chesapeake Bay. *Coastal Management* 22:119–145.

Holling, C. S., ed. 1978. *Adaptive Environmental Assessment and Management.* John Wiley, New York, NY.

———. ed. 1979. Resilience and stability of ecological systems. *Annual Review of Ecology and Systematics* 41:1–23.

Horton, Tom. 1991. *Turning the Tide: Saving the Chesapeake Bay.* Island Press, Washington, DC.

Hutter, M.A. 1985. The Chesapeake Bay: Saving a national resource through multi-state cooperation. *Virginia Journal of Natural Resources Law* 4(2):185–208.

Imperial, M. T., D. D. Robadue, Jr., and T. Hennessey. 1992. Managing coastal and estuary environmental quality in the United States: An evolutionary perspective on the development of the National Estuary Program. *Coastal Management* 20:311–341.

———. 1993. The evolution of adaptive management for estuarine ecosystems. The National Estuary Program and its precursors. *Oceans and Coastal Management* 20:147–180.

Kiser, L. and E. Ostrom. 1982. The three worlds of action. In E. Ostrom, *Strategies of Political Inquiry.* Sage Publications, Beverly Hills, CA.

Lee, K. 1993. *Compass and Gyroscope: Integrating Science and Politics for the Environment.* Island Press, Washington, DC.

Lee, K. and J. Lawrence. 1986. Restoration under the Northwest Power Act. *Environmental Law* 16:423–459.

Oakerson, R. 1986. A model for the analysis of common property problems. In *A Conference on Common Property Resources.* National Research Council, National Academy Press, Washington, DC.

Odum, E. 1983. *Basic Ecology.* J.B. Saunders, Philadelphia, PA.

Ophuls, W. 1977. *Ecology and the Politics of Scarcity.* W.H. Freeman, San Francisco, CA.

Ophuls, W. and S. Boyan. 1992. *Ecology and the Politics of Scarcity Revisited.* W.H. Freeman, San Francisco, CA.

Ostrom, E. 1990. *Governing The Commons.* Cambridge University, Cambridge, UK.

Resources for the Future. 1979. *An Evaluation of Institutional Arrangements for the Chesapeake Bay.* Washington, D.C.

Schobel, J. 1981. *The Living Chesapeake.* Johns Hopkins Press, Baltimore, MD.

Schubel, J. R. 1986. *The Life and Death of the Chesapeake Bay.* University of Maryland, College Park, MD.

Sproule-Jones, M. 1980. *The Real World of Pollution Control.* University of British Columbia Press, Vancouver, BC.

Tang, Shui Yam. 1992. *Institutions for Collective Action.* ICS Press, San Francisco, CA.

Tripp, J. T. B. and M. Oppenheimer. 1988. Restoration of the Chesapeake Bay: A multi-state institutional challenge. *Maryland Law Review* 47:425–450.

U.S. Environmental Protection Agency. Chesapeake Bay Program. 1981. *Chesapeake Bay: An Introduction to an Ecosystem.* Annapolis, MD.

———— . 1983a. *Chesapeake Bay: A Framework for Action.* Annapolis, MD.

———— . 1983b. *Chesapeake Bay Program: Findings and Recommendations.* Philadelphia, PA.

———— . 1983c. *Final Report and Recommendations: Resource Users Management Team.* Philadelphia, PA.

———— . 1990. *Report and Recommendations of the Non-Point Source Evaluation Panel.* Annapolis, MD.

U.S. Senate. Committee on Governmental Affairs. 1985. *Roundtable Discussion of the Federal Role in the Chesapeake Bay Cleanup Program.* Washington, DC.

Walters, C. 1986. *Adaptive Management and Renewable Resources.* Macmillan, New York, NY.

Warner, J. W. and J. Warren. 1985. Land based pollution and the Chesapeake Bay. *Washington and Lee Law Review* 42:1099–1138.

Williamson, Oliver. 1975. *Markets and Hierarchies.* Free Press, New York, NY.

Note

1. This paper is drawn from a comparative study of the governance of estuaries. In addition to Chesapeake, our study covered the following: Puget Sound and Narragansett, Delaware, and Galveston Bays. We did additional work on the National Estuary Program. Research for the comparative study was partially supported by The National Oceanic and Atmospheric Administration Office of Sea Grant, U.S. Department of Commerce under Grant No. NA-89-AA-D-SG-082. For recent publications from the study see Hennessey (1994); Healey and Hennessey (1994); Imperial, Robadue, and Hennessey (1993); and Imperial, Robadue, and Hennessey (1992).

CHAPTER TEN

Ecosystem Functions and Ecosystem Values

Dennis F. Whigham

Introduction

Our understanding of how ecosystems function has expanded enormously since the term was first defined as a *dynamical system consisting of a biological entity, typically a regional biota (community), together with its environment* by Alfred George Tansley in 1935. In his analysis of the history of the ecosystem concept, Golley (1993) concluded that "the ecosystem, for some at least, has provided a basis for moving beyond strictly scientific questions to deeper questions of how humans should live with each other and the environment. In that sense, the ecosystem concept continues to grow and develop as it serves a larger purpose." Indeed, many societal questions related to the environment are ecosystem related (Odum 1989; Cairns et al. 1992). Here, I emphasize the ongoing national debate about wetlands, such as those that occur in the Chesapeake Bay watershed, to demonstrate the utility of the ecosystem concept. I begin with an overview of the ecosystem concept and how it is being applied when decisions have been made about the use, restoration, or conservation of natural resources. I then discuss the importance of separating ecological "functions" from societal "values" when applying the ecosystem concept. I conclude the discussion with an example of a method that is being developed to provide ecological input (i.e., information on wetland functions) to individuals who make decisions about the fate of wetlands.

The Ecosystem Concept and Society

I have chosen two examples. Chesapeake Bay and South Florida, to demonstrate the current centrality of the ecosystem concept in solving societal problems. The health of Chesapeake Bay has been of central concern for more than a decade (Horton and Eichbaum 1991; Boynton, this volume), and many of the undesirable ecological and economic changes that have occurred over the past century have been associated with human activities (Brush, this volume). Fisheries resources have declined, beds of submersed aquatic vegetation have almost disappeared, and water quality has deteriorated (Dennison et al. 1993). The causes of the negative trends in the biotic and abiotic resources of the Chesapeake Bay are numerous (e.g., wetland losses, over-exploitation, disease, increased sediment, and nutrient inputs). Finding solutions to these problems individually and collectively has proven difficult, but almost everyone recognizes that management activities must focus on the entire watershed and not just the estuary. Thus, there is clear recognition that the entire ecosystem, the Chesapeake Bay and its watershed, must be managed if it is to continue to provide the goods (e.g., shellfish) and services (e.g., improvement of water quality) that have historically been associated with it. One management goal, for example, is to reduce the amount of nonpoint runoff within the watershed by 40 percent. Attaining that goal will be difficult, and it will require a range of activities including the use of Best Management Practices on farms, the restoration of riparian habitats along streams, and the creation/restoration of wetlands to intercept nutrients and sediments before they reach streams and, ultimately, the Chesapeake Bay. An analogous situation exists in South Florida.

Water flow through much of South Florida is controlled by a complex network of levees, water storage areas, channels, and large-scale pumping that has been developed over the past 90 years (Light and Dineen 1994). Most of the water control activities were designed to control flooding and manage water flow to permit regional urban/suburban and agricultural development, but consequences include a dramatic reduction in the size of the original Everglades ecosystem (Frayer and Hefner 1991; Davis and Ogden 1994); political and legal battles that primarily focused on changes in the hydrologic and phosphorus inputs into Everglades National Park (Davis 1994); and the restoration of the Kissimmee River (Dahm 1995). Similar to the Chesapeake Bay example, restoration and conservation in South Florida will require an ecosystem approach no matter whether the goal of a particular activity is the conservation of a species (e.g., Florida Panther) or the conservation of a large part (e.g., Everglades National Park) of the landscape (Gunderson et al. 1995). Davis and Ogden (1994) summarize a series of papers related to restoration of the Everglades, and each of the following recommendations involves ecosystem management:

- Their first recommendation focuses on the importance of spatial scale in ecosystems. They recommend that "the reduction in ecosystem size and compartmentalization of the remaining system are trends that must be reversed in any Everglades restoration initiative."
- The second recommendation considers disturbance and is based on the concept that when ecosystems are of sufficient spatial scale, there is adequate space to offset local disruptive effects and that disturbances contribute to the overall diversity of the system. They recommend that environmental fluctuations and extremes in hydrology and fire proceed as they would have in the natural Everglades system.
- Hydrologic parameters are key to the normal functioning of wetland ecosystems. Rainfall is the source of water for the South Florida system, and Davis and Ogden recommend that water delivery plans be developed that are based upon antecedent rainfall for all major areas of remnant Everglades marshland, including the Water Conservation Areas.
- In addition to the amount of water, the pattern of flow and distribution of water in time and space are important. They recommend to "incorporate components into rainfall-based water delivery plans that will restore flow volumes and distributions in time and space, as simulated by natural system hydrology model."
- The spatial and temporal pattern of water depth is important to food webs, reproduction, and survival of many species of flora and fauna in the Everglades. They recom-

mend to "build components into rainfall-based water delivery plans that will restore depth patterns in time and space, as simulated by the natural system hydrology model."

- Finally, prolonged hydroperiods are essential for the maintenance of the peat-based wetlands within the Everglades system. They recommend that rainfall-based water delivery plans be developed to mimic extended periods of flooding as would have occurred in the remnant Everglades marshes under predrainage conditions."

There are numerous other examples of how the ecosystem concept has become central to conservation and restoration activities. Indeed, almost all issues related to endangered species, biodiversity, and environmental conservation ultimately must focus on management of the ecosystem(s) in which the species are found (Gillis 1990; Franklin 1995; Willcox 1995). Unfortunately, the ecosystem concept has not made substantial contributions to other fields of endeavor (Peters 1991; Lovejoy 1995), and decisions that influence ecosystems are regularly made by individuals who know little about ecosystem structure and function (Franklin 1995). A good example comes from the current national debate about wetlands.

The Confusion Between Wetland Functions and Values

What is a wetland? If we did not have to draw lines on maps to show where wetlands occur, this question would not be debated at all and just about everyone would be satisfied with the science-based definition recently recommended by the Committee on Wetlands Characterization of the National Academy of Sciences (NAS 1995). The Committee defined a wetland as an

"*ecosystem* that depends on constant or recurrent, shallow inundation or saturation at or near the surface or the substrate. The minimum essential characteristics of a wetland are recurrent, sustained inundation or saturated at or near the surface and the presence of physical, chemical, and biological features reflective of recurrent, sustained inundation or saturation. Common diagnostic features of wetlands are hydric soils and hydrophytic vegetation. These features will be present except where specific physiochemical, biotic or anthropogenic factors have removed them or prevented their development."

This definition is not that different from others that have been widely accepted or used in the field of wetland science (Mitsch and Gosselink 1993). The NAS definition was, however, ignored by the U.S. House of Representatives in passing H.R. 961, the 1995 reauthorization of the Clean Water Act.

H.R. 961 has several provisions and definitions that are not justified scientifically, nor are they consistent with the NAS wetland definition (Zedler 1995). For example, H.R. 961 provides for and requires the characterization of all U.S. wetlands into three categories:

- Type A wetlands would be those that have the highest *functional value* and are of critical significance to the long-term conservation of that type of wetland *ecosystem;*
- Type B wetlands would be those that provide *significant wetland functions;*
- Type C wetlands are those that provide *limited wetland function.*

There is no scientific basis for these three categories. Instead, the use of terms such as *functional value, significant wetland functions, limited wetland function* are subjective at best and defined by the dominant political view (or agenda) of the day. This subjectivity is underscored by H.R. 961 itself, which requires that no county or parish can have more than 20 percent of its area defined as Type A wetlands. Wetlands that provide important goods and services in support of human activities can and do exceed this areal extent in some regions of the country (Figure 10–1). In the Maryland counties of Dorchester and Somerset, for example, more than 35 percent of the total land area of each is wetlands, and most would be classified as Type A. The surface waters from both counties drain into Chesapeake Bay, and their economies rely heavily on resources from the Chesapeake Bay. Why should Dorchester and Somerset Counties be treated differently than the other 21 Maryland counties because their wetlands are more extensive?

I have italicized the word *ecosystem* in the NAS definition and several words (*functional value, ecosystem, significant wetland functions, limited wetland functions*) related to H.R. 961. It is significant that both characterizations of wetlands include the word

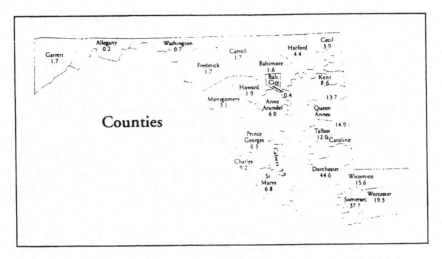

Figure 10–1. Percentage of total land area in wetlands for countries in Maryland.
Source: Tiner and Burke 1995.

ecosystem because it further demonstrates that the ecosystem concept has been widely recognized in both scientific and political arenas. The words *functional value* and *functions* that appear in H.R. 961 show, however, that ecologists have not done a very good job of separating the issue of *what ecosystems do* from the reality that *human societies receive goods and services from ecosystems* that have economic value (van Wilgen et al. 1996). Another interpretation would be that ecologists often have not done a good job of promoting value(s). Ecologists can demonstrate functions but have not adequately assigned values that sway public/political opinions. What have ecologists learned from the study of ecosystems that needs to be communicated to individuals who strive to make environmental decisions that serve the public good?

Perhaps most importantly, we have learned that ecosystems are more than the sums of their parts and that they have emergent properties such as ecosystem stability and ecosystem diversity (Odum 1989; Golley 1993). Second, ecologists have learned that humans are integral components of ecosystems and that there is no place on the face of planet Earth that is not or has not been influenced in one way or another by human activities. Almost all human activities impact ecosystems and their constituent species. What we have not fully realized nor appreciated is that systems of economics are linked in a variety of ways to natural systems (Freeman, this volume; Peskin, this volume).

Van Wilgen et al. (1996) show the linkages between "ecosystem services" and economic values for fynbos (shrubland) watersheds in arid South Africa. Watersheds dominated by native fynbos species deliver cleaner water and more water than watersheds dominated by alien trees and shrubs. Economically, the unit cost for a m^3 of water is 11.9 cents for watersheds dominated by native fynbos, compared to 13.8 cents per m^3 of water for watersheds that were dominated by alien plants which require management. Howard Odum, a long-time proponent of the unity of natural systems and economic systems, has suggested that because of commonalities of systems that there are many similarities between economic systems and ecological systems (Odum 1983). Materials and energy flows in natural ecological systems (Odum 1989) and in economic systems, money flows in response to the flow of goods and services but in the opposite direction of material and energy flow. In the fynbos example described above, money that is required for management of alien flora flows to individuals who perform the work. In response to the flow of money, a return on investment is realized in the form of increased water flow and improved water quality. Odum suggests that it is possible to develop models to understand the relationships between ecosystems and economic systems using a common currency that links the flows of goods and services to the natural services (i.e. functions) of ecosystems. If Odum's suggestions are correct, why then has it been so difficult to relate economic systems to ecological systems (e.g., Bender et al. 1994)?

In part, our inability to relate economic and ecological systems has resulted from confusion over ecosystem functions and ecosystem values. Functions are the normal characteristic actions or activities of something (Smith et al. in press). In an ecological context, functions are the processes that are necessary for the self-maintenance of ecosystems. Values are the rules that determine what people consider important. Brown (1984) differentiated between held and assigned values. Held values are the precepts in which individuals or groups believe. Assigned values are the indicators of the relative importance of something to an individual or group. In an ecological context, functions exist in the absence of human activities and are a normal part of the self-sustaining properties of ecosystems. Values are the goods and

services that emanate from functions (Taylor et al. 1990). The relationship between functions and values in the context of wetland ecosystems is shown in Figure 10–2. Table 10–1 (Smith et
al. in press) and Richardson (1994) further show the relationships between wetland functions and values.

The dashed line in Figure 10–2 separates the physical (e.g.
geomorphic) setting where wetlands occur in the landscape from
societal interactions with wetlands. Items above the dashed line
can continue in the absence of human activities (i.e., they are
wetland functions) while those below the line show how wetland
products and functions are used (i.e. wetland values). Critical
processes and mechanisms such as photosynthesis and ecosystem functions such as primary production and biomass accumulation may become resources for human life support. The

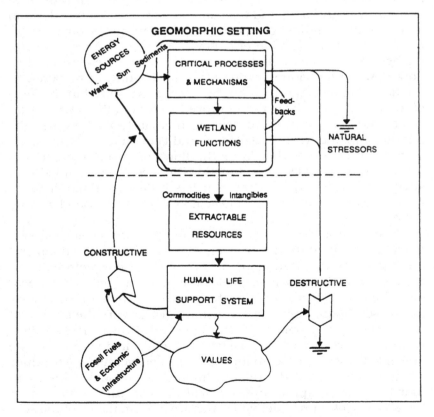

Figure 10–2.   The relationship between functions and values in the context of ecosystems. Source: Brinson 1993.

Table 10–1. Wetland functions and the values (i.e., benefits, goods, and services) that they provide (Smith et al. in press).

FUNCTIONS	VALUES
Store Surface Water:	
The ability of a wetland to temporarily store surface water arriving as precipitation, overland or subsurface flow from adjacent upland areas, or originating from upstream or downstream (e.g. tidal) areas and arriving via overbank flooding.	Reduces flood related damage downstream
Reduce the Energy Level of Surface Water:	
The ability of a wetland to reduce the energy of moving surface water due to structural roughness in the wetland.	Reduce erosion from storms and floodwaters
Recharge Ground Water:	
The ability of a wetland to provide a conduit for the recharge of a ground water aquiter.	Maintain pumpable supplies of ground water.
Discharge Ground Water:	
The ability of a wetland to provide a conduit for the discharge of ground water.	Maintain stream and lake water levels
Stabilize Soils:	
The ability of a wetland to protect soil from the erosive action of currents and waves because of presence of litter and the binding action of roots.	Reduce erosion of shorelines and streambanks from storms and floods
Detain, Remove, and Transform Nutrients:	
The ability of a wetland to temporarily detain, or permanently remove nutrients from surface or ground water column through biochemical transformation, incorporation into biomass, or burial.	Maintain surface and ground water quality
Detain, Remove, and Transform Contaminants:	
The ability of a wetland to temporarily detain, or permanently remove contaminants from the water column through incorporation into biomass, biogeochemical transformation, or burial.	Maintain surface and ground water quality

(Continued)

Table 10–1.   Wetland functions and the values (i.e., benefits, goods, and services) that they provide (Smith et al. in press). (Continued)

Detain and Remove Sediments:	
The ability of a wetland to temporarily detain, or permanently remove suspended sediments from surface water through sedimentation and burial.	Maintain surface water quality
Provide Ecosystem, Landscape, and Global Integrity	
The ability of a wetland to provide the conditions required for the biotic and abiotic processes characteristic of the wetland ecosystem, its landscape, and the world.	Maintain ecosystem, landscape, and global processes
Provide Wetland Ecosystem Structure:	
The ability of a wetland to provide the unique structural characteristics of wetland ecosystems.	Maintain populations of wetland dependent plant and animals species, preserve endangered species, maintain biodiversity, provide dispersal corridors

term extractable resources is meant to include intangibles, commodities, and all other goods and services that contribute to the human's life support system. A key to understanding the relationships between functions and values is provided in the feedback mechanisms which are called CONSTRUCTIVE and DESTRUCTIVE in Figure 10–2. Ecosystems can only continue to provide human goods and services when the feedback mechanisms allow ecosystem functions to continue at a sustainable level. When ecosystem functions cease or change due to overexploitation (i.e. destructive feedbacks), the goods and services to human societies cease to be provided. The linkages between ecosystem functions and goods and services are central to the concept of sustainability and a key concept that ecologists need to describe adequately to individuals who are involved in ecosystem management.

Wetland Assessment— Functions Versus Values

The national wetland debate has primarily focused on such lofty issues as wetland definitions, delineation, and wetland classification (NAS 1995). Local wetland issues, in contrast, often deal with the fate of a specific piece of real estate and questions such as: can a developer or homeowner fill a small section of wetland on his/her property; can the State Department of Transportation use fill to place a road across a floodplain, etc.? Decisions about the fate of wetlands have been made in a variety of ways ranging from decisions based on best professional judgment to decisions based on the application of complex functional assessment methods.

Over the last 20 years, several methods have been developed to perform functional assessment of wetlands (Larson and Mazzarese 1994; NAS 1995), the goals of which have been to develop reproducible procedures to assess the functions of wetlands. One of the most widely cited and applied assessment methods (FHWA) was developed by the Federal Highway Administration (Adamus and Stockwell 1983). FHWA was later modified into what become known as the Wetland Evaluation Technique (WET), which included an interactive computer analysis program (Adamus 1987; Adamus et al. 1991). WET, which was further developed into WET 2.0 by the U.S. Army Corps of Engineers, was the first widely used method to assess all recognized wetland functions. WET and WET 2.0, however, did not adequately separate wetland functions from wetland values; an important distinction since "the key ecological question that needs to be addressed under any development scenario is whether or not wetland functions have been significantly altered" (Richardson 1994).

Smith et al. 1995 describe the rationale behind a hydrogeomorphic (HGM) approach to a functional assessment system that is being developed for wetlands in the U.S. (Brinson 1993). The first national HGM models have been developed for riverine wetlands (Brinson et al. 1995) and specific regional models are currently being tested and evaluated. The HGM models differ from previous assessment models in three ways. First, HGM recognizes that the set of functions that need to be assessed are likely to differ between wetland types. Second, it uses functional indices that can be quantified and the quantitative scales are based on ecological data gathered from reference wetlands. Third, the developers of HGM have attempted to limit the models to wetland functions and not include wetland values.

Table 10–1 defines 10 HGM functions of riverine wetlands (Smith et al. in press) and gives examples of the values that emanate from the ecological processes. The first four functions (store surface water; reduce the energy level of surface water; recharge ground water; discharge ground water) are related to the movement of water into and through wetlands and result in widely recognized values (i.e., reduce flood damage and erosion, provide supplies of ground water). Four functions (stabilize soils; detain, remove and transform nutrients; detain, remove and transform contaminants; detain and remove sediments) are related to the cycling and retention of nutrients and contaminants which are associated with improved water quality. The last two functions (provide ecosystem, landscape, and global integrity; provide wetland ecosystem structure) are related to processes that maintain flora and fauna in the wetland and in landscapes that contain wetlands. Values that result from healthy wetlands include recreational values associated with wetlands as well as the extraction of resources (i.e., animal pelts, fish, waterfowl) from wetlands and associated water bodies.

In summary, I have attempted to document the importance of the ecosystem concept and demonstrate that ecosystems provide numerous goods and services to society which can only continue to be benefits if healthy ecosystems are maintained. I have described differences between ecological processes and ecosystem values to demonstrate that ecosystem management will have a higher probability of success when the two (i.e., functions and values) are kept separate. I believe that it is imperative to delineate clearly the functions of wetlands as a first step to making social and political decisions. Only through providing and implementing the types of formal functional assessments described above can decision makers (resource managers, state and federal agency personnel, politicians) truly characterize mechanisms and processes and give them appropriate weight in arriving at well-educated and balanced value assessments. This is essential to our sustained ability to continue to share the earth with other organisms and benefit from the goods and services that we enjoy from ecosystems.

Acknowledgments

I would like to thank all of my colleagues who participated in the development of the functional assessment methodology (HGM) described in this paper. I am especially thankful to Bill Ainsle, Mark Brinson, Garry Hollands, Lyndon Lee, Dennis

McGee, Dick Novitski, Wade Nutter, Rick Hauer, Rick Rheinhardt, Jeff Mason, Mark Rains, and Dan Smith for the many stimulating discussions that were held during the development of the conceptual framework for HGM. The manuscript was greatly improved by comments and suggestions from Greg Ruiz, Norm Christensen, and David Simpson.

References

Adamus, P. R. 1987. *Wetland Evaluation Technique (WET): Volume II-Methodology*. U.S. Army Corps of Engineers, Waterways Experiment Station, Vicksburg, MS.

Adamus, P. . and L. T. Stockwell. 1983. *A Method for Wetland Functional Assessment, Volume I*. Office of Research and Development, Federal Highway Administration, U.S. Department of Transportation, Washington, DC.

Adamus, P. R., L. T. Stockwell, F. J. Clarain, Jr., M. E. Morrow, L. E. Rozas, and R. D. Smith. 1991. *Wetland Evaluation Technique (WET), Volume I: Literature Review and Evaluation Rationale*. Wetlands Research Program Technical Report WRP-DE-2. U.S. Army Corps of Engineers, Waterways Experiment Station, Vicksburg, MS.

Bender, M. J., G. V. Johnson, and S. P. Simonovic. 1994. Sustainable management of renewable resources: a comparison of alterative decision approaches. *International Journal of Sustainable Development and World Ecology* 1:77–88.

Brinson, M. M. 1993. *A Hydrogeomorphic Classification for Wetlands*. U.S. Army Corps of Engineers. Wetlands Research Program. Technical Report WRP-DE-4. Washington, DC.

Brinson, M. M., W. Kruczynski, L. C. Lee, W. L. Nutter, R. D. Smith, and D. F. Whigham. 1994. Developing an approach for assessing the functions of wetlands. pp. 615–624. In W.J. Mitsch (ed.), *Global Wetlands: Old World and New*. Elsevier Science B.V., Amsterdam, The Netherlands.

Brinson, M. M., F.R. Hauer, L. C. Lee, W. L. Nutter, R. D. Rheinhardt, R. D. Smith, D. F. Whigham. 1995. *A Guidebook for Application of Hydrogeomorphic Assessments to Riverine Wetlands*. Wetlands Research Program Technical Report WRP-DE-11. U.S. Army Corps of Engineers, Waterways Experiment Station, Vicksburg, MS.

Brown, T. C. 1984. The concept of value in resource allocation. *Land Economics 60:* 231–246.

Cairns, John Jr., B. R. Niederlehner, and D. R. Orvos. 1992. *Predicting Ecosystem Risk.* Princeton Scientific Publishing Co., Inc., Princeton, NJ.

Dahm, C. N. (Guest editor). 1995. Special Issue: Kissimmee River Restoration. *Restoration Ecology* 3(3):145–238.

Davis, S. 1994. Phosphorus inputs and vegetation sensitivity in the Everglades. pp. 357–378. In S. Davis and J. Ogden (eds.), *Everglades: The Ecosystem and its Restoration.* St. Lucie Press, Delray Beach, FL.

Davis, S. and J. Ogden (eds.). 1994. *Everglades: The Ecosystem and Its Restoration.* St. Lucie Press, Delray Beach, FL.

Dennison, W. C., R. J. Orth, K. A. Moore, J. C. Stevenson, V. Carter, S. Kollar, P. W. Bergstrom, and R. A. Batiuk. 1993. Assessing water quality with submersed aquatic vegetation. *BioScience* 43:86–94.

Franklin, J. F. 1995. Scientists in wonderland. *BioScience* Supplement 74–78.

Frayer, W. E. and J. M. Hefner. 1991. *Florida Wetlands. Status and Trends, 1970s to 1980s.* U.S. Fish and Wildlife Service, Atlanta, GA.

Gillis, A. M. 1990. The new forestry. *BioScience* 40:558–562.

Golley, F. B. 1993. *A History of the Ecosystem Concept in Ecology.* Yale University Press, New Haven, CT.

Gunderson, L. H., S.S. Light, and C.S. Holling. 1995. Lessons from the Everglades. *BioScience* Supplement 66–73.

Horton, T. and W. M. Eichbaum. 1991. *Turning the Tide: Saving the Chesapeake Bay.* Island Press, Washington, DC.

Larson, J. S. and D. B. Mazzarese. 1994. Rapid assessment of wetlands: History and application to management. pp. 625–636. In W.J. Mitsch (ed.), *Global Wetlands: Old World and New.* Elsevier Science B.V., Amsterdam, The Netherlands.

Light, S. S. and J. W. Dineen. 1994. Water control in the Everglades: A historical perspective. pp. 47–83. In. S. Davis and J. Ogden (eds.), *Everglades: The Ecosystem and its Restoration.* St. Lucie Press, Delray Beach, FL.

Lovejoy, T. E. 1995. Will expectedly the top blow off? *BioScience* Supplement 3–6.

Mitsch, W. J. and J. G. Gosselink. 1993. *Wetlands,* 2nd edition. Van Nostrand Reinhold Company Inc., New York, NY.

National Academy of Sciences. 1995. *Wetlands: Characteristics and Boundaries,* National Academy Press, Washington, DC.

Odum, E. P. 1989. *Ecology and Our Endangered Life-Support Systems.* Sinauer Associates, Inc., Sunderland, MA.

Odum, H. T. 1983. *Systems Ecology.* John Wiley & Sons, New York, NY.

Peters, R. H. 1991. *A Critique for Ecology.* Cambridge University Press, New York, NY.

Richardson, C. J. 1994. Ecological functions and human values in wetlands: a framework for assessing forestry impacts. *Wetlands* 14:1–9.

Smith, R. D., A. Ammann, C. Bartoldus, P. Garrett, and M. M. Brinson. In press. *An approach for assessing wetland functions using Hydrogeomorphic Classification reference wetlands, and functional indices.* Wetlands Research Program, U.S. Army Corps of Engineers Waterways Experiment Station, Technical Report WRP-DE-9. Vicksburg, MS.

Taylor, J. R., M. A. Cardamone, W. J. Mitsch. 1990. Bottomland hardwood forests: Their functions and values. pp. 13–88. In. J. G. Gosselink, L. C. Lee, and T. A. Muir, (eds.), *Ecological Processes and Cumulative Impacts: Illustrated by Bottomland Hardwood Ecosystems.* Lewis Publishers, Chelsea, MI.

Tiner, R. W. and D. G. Burke. 1995. *Wetlands of Maryland.* U.S. Fish and Wildlife Service, Ecological Services, Region 5. Hadley, MA.

van Wilgen, B., R. M. Cowling, and C. J. Burgers. 1996. Valuation of ecosystem services. *Bioscience* 46:184–189.

Willcox, L. 1995. The Yellowstone experience. *BioScience* Supplement 79–83.

Zedler, J. B. 1995. Reinventing wetland science. *National Wetlands Newsletter* 17:1, 17–20.

CHAPTER ELEVEN

On Valuing the Services and Functions of Ecosystems

A. Myrick Freeman III

Introduction

One of the impediments to making progress on the task of valuing ecosystems is the various meanings economists and natural scientists attach to terms such as value, ecosystem functions, and ecosystem services. The discussions at the workshop were quite useful, not because they resolved issues and problems of terminology (they did not), but because they helped us to understand more clearly what these issues and problems are.

I have found that economists and ecologists typically use the term "value" (either as a noun or verb) in two different senses when they use it in discussions of ecosystems. Ecologists usually use the term to mean "that which is desirable or worthy of esteem for its own sake; thing or quality having intrinsic worth."[1] Economists use the term in a sense more akin to "a fair or proper equivalent in money, commodities, etc . . . ," where "equivalent in money" represents that sum of money which would have an equivalent effect on the welfare or utilities of individuals. Toman (this volume) also alludes to this distinction when he describes economists' concept of value as anthropocentric and utilitarian and says that critics of the economic approach to valuation charge that it devotes "insufficient attention to the intrinsic worth of ecosystems. . . . "

This difference in focus or perspective on value also carries over to the distinction between ecosystem functions from an ecological point of view and ecosystem services from an economic point of view. Ecosystem functions include such things as primary productivity (photosynthesis), decomposition, nutrient recycling, etc. Ecosystem service flows are the materials and services provided by ecosystems that enhance human welfare and therefore are valued by people. Examples of service flows include such things as wood and fiber from forests, and amenities such as scenic vistas, hiking, and wildlife observation. Ecosystem functions may have intrinsic worth or value in the eyes of ecologists and other observers. But they do not have economic value unless they help to support service flows to people.

During the first morning of the workshop, Jerry Franklin said that ecosystem functions are the work that an ecosystem does. I think that this is a useful analogy. In engineering and physics, not all of the work performed by a system is useful work. For example, falling water performs useful work when it passes through a turbine connected to an electrical generator. But without the turbine, the work performed is not useful to people and has no economic value. Similarly, the photosynthesis carried out by a rice plant is useful, at least that part of it that is necessary for the production of grain.[2] But if a change in photosynthesis has no effect on any of the service flows of the system to people, then it does not represent a change in the ecosystem's useful work. And there is no economic value connected with that change.

Sometimes the connection between an ecosystem function and an economically valuable service flow maybe quite direct, as in the case of photosynthesis producing useful plant material. But in many cases, the connection can be indirect and quite subtle. For example, photosynthesis by wildflowers may help to support a population of wild bees that also pollinate commercially valuable fruits. Other examples of more indirect linkages will be provided in the final section of this paper.

What I am suggesting here is that the key to valuing a change in an ecosystem function is establishing the link between that function and some service flow valued by people. If that link can be established, then the economist's concept of derived demand can be applied. The value of a change in an ecosystem function can be derived from the change in the value of the ecosystem service flows it supports. For example, in the economic theory of production, the value of one more hour of labor services (work) is the value of the additional output that the work makes possible (its value of marginal product). Similarly, if an increase in the rate of photosynthesis in an ecosystem results in an increase in the flow of economically valuable food or fiber, the economic value of the increase in photosynthesis is the increase in the value of food or fiber that it supports.[3]

The next section provides a more detailed explanation of the economic perspective on ecosystems as sources of valued services to people. I then describe four types of service flows for which estimates of economic values have been made and explain why these efforts fall short of the valuation of ecosystem functions. Finally, I suggest some situations in which it may be possible to make the link between an ecosystem function and a service flow that has economic value.

Ecosystems and Their Service Flows

Many economists have come to realize that it can be useful to think of natural resource systems such as forests, estuaries, and air sheds as natural assets that yield flows of valuable services to people (Smith 1988; Gottfried 1992; Freeman 1993b). A forest such as a unit in the U.S. National Forest System is a good example of a resource-environmental system that provides a wide range of services including materials such as wood and fiber and the amenities associated with a variety of outdoor recreation activities. The forest system also regulates stream flow, controls erosion, and absorbs atmospheric carbon dioxide.

An estuary and its adjacent upland provide services such as support for commercially exploitable fisheries, space for residential, industrial, and commercial structures, and the absorption of waste products from local runoff and from upstream sources along rivers and streams. The estuary system also provides a variety of amenity services associated with recreation activities (Freeman 1995a).

Airsheds provide life support and climatic regulatory services, and amenity services in the form of clear views of distant mountains, and they absorb and disperse waste products from chimneys and exhaust pipes.

Some economists have suggested that the service flows from these natural resource systems can be grouped into four categories (Freeman et al. 1973; Freeman, 1993b). First, resource systems serve as sources of material inputs to the economy such as fossil fuels, wood products, minerals, water, and fish. Second, some components of resource systems provide life support services in the form of a breathable atmosphere and a liveable climatic regime. Third, resource systems provide a wide variety of amenity services, including opportunities for recreation, wildlife observation, the pleasures of scenic views, and perhaps even services that are not related to any direct use of the environment. These latter are sometimes called nonuse or existence values. Finally, resource systems can disperse, transform, and store the waste products that are generated by economic activity and discharged into the environment.

The values of these service flows derive from their contribution to human well-being. Economists have developed a framework for defining the values of changes in these service flows in terms of the sum of money or purchasing power that would have an equivalent effect on human well-being.[4] This framework is based on the assumption that people have well-defined preferences among alternative bundles of goods and services, including nonmarket goods and environmental services. It also assumes that people know their preferences and that these preferences have the property of substitutability among the market and nonmarket goods making up the bundles. By substitutability, economists mean that if the quantity of one good in an individual's bundle is reduced, it is possible to increase the quantity of some other good so as to leave the individual no worse off because of the change. In other words, the increase in the quantity of the second good substitutes for the decrease in the first good. The property of substitutability is at the core of the economist's concept of value because substitutability establishes tradeoff ratios between pairs of goods that matter to people.

The tradeoffs that people make as they choose less of one good and substitute more of another reveal something about the values people place on these goods. If one of the goods has a monetary value, the revealed values are monetary values. The money price of a market good is just a special case of a tradeoff ratio because the money given up to purchase one unit of one element of the bundle is a proxy for the quantities of one or more of the other elements in the bundle that had to be reduced in order to make the purchase.

The value measures based on substitutability can be expressed either in terms of willingness to pay (WTP) or willingness to accept compensation (WAC). WTP and WAC measures can be defined in terms of any other good that the individual was willing to substitute for the good being valued. Money is simply the most convenient numeraire for expressing tradeoff ratios. WTP and WAC could be measured in terms of any other good that mattered to the individual.

From an economic perspective, ecosystems are a form of natural asset that provide service flows to people, and their economic values derive from the contributions that these service flows make to human welfare.[5] The economic value of an ecosystem as a type of natural asset is the sum of the discounted present values of the flows of all the services it provides to people. Since many of these service flows are not bought or sold in markets and therefore do not have market prices, the economic value of an ecosystem may be quite different from its market value. For example, an acre of wetland might trade in the land market on the basis of its value for commercial or residential development. But this value could be quite different from the value of its services as wildlife habitat and as a means of controlling floods and recharging ground water aquifers.

There may be circumstances in which it is important to know the total economic value of an ecosystem. But most policy questions involve not the complete elimination of an ecosystem, but rather changes in the level of one or more of its service flows. The analytical tools of economics are well suited to dealing with the values of such incremental or marginal changes. A number of economic methods have been developed for these purposes. Toman (this volume) briefly describes three of the major types of methods. They are: the travel cost model of recreation demand in which the time and money cost of traveling to a recreation site is treated as an implicit price; the hedonic price method in which differences in prices of products are related to differences in some qualitative attribute, for example the level of air quality associated with one's residence; and contingent valuation, where people are asked questions about their willingness to pay or to make trade-offs for changes in environmental services.[6] Most of the applications of these methods have focused on changes in some measure of environmental quality. Examples include the valuation of the effects of changes in air pollution on human health, reductions in visibility associated with air pollution, and changes in recreational opportunities linked to changes in water pollution levels. Only occasionally have these methods been applied to valuing some aspect of ecosystem functions or service flows. The next section describes four groups of studies that focus on ecosystems.

Valuing Changes in Ecosystem Service Flows

In order to estimate the economic value of a basic ecosystem function, we need to know the link between that function and the ecosystem service flows that it supports, and we need to know the economic value of that service flow: that is, people's willingness to pay. In the four types of studies described in this section, measures of the economic value of the service flows were available, either from market data or from the application of nonmarket valuation methods. In two of the cases, there was also some information available on the underlying ecosystem circumstances, but not enough to link the service flows to a specific ecosystem function. And in the other two cases, there was no information on the link between the service flow and the ecological system.

Wetlands Productivity

The type of study that comes closest to attributing economic values to ecosystem functions involves linking commercial and recreational fishing activities to the supply of tidal wetlands that support the populations of the exploited fish species. In two applications of this approach, time series data on wetland acreage, the level of fishing effort, and other control variables were used in regression equations to explain a physical measure of output or service flow. In the first study, the output was the harvest of blue crabs along the Florida Gulf Coast (Lynne, Conroy, and Prochaska 1981). In the second study, separate equations were estimated for the recreational and commercial harvests of eight fish species on the Florida Gulf Coast (Bell 1989). Both studies found that changes in tidal wetland acreage were associated positively with changes in the rates of harvest of certain species after controlling for other factors. For commercial harvests, the marginal economic value of additional wetland acreage was calculated by using the net monetary value per unit of harvest to convert the change in physical yield into an economic value. For recreational fishing, the economic value was based on estimates of individuals' increased willingness to pay for access to the more successful fishing activity.

There are two observations to make about these two studies. First, although the studies provide estimates of the economic value of an acre of tidal wetland, they do not provide values of any specific ecosystem function. Rather, they provide estimates of the contribution of all of the ecosystem functions carried out on an acre of wetland to the fish species analyzed. Second, these studies provide values of the contribution of these functions to only one service flow, the service flow embodied in the commercial and recreational harvest of these fish species. The values of other service flows and the ecosystem functions that support them are not captured by these studies.

Agricultural Productivity

The second group is a series of studies of the economic consequences of ozone air pollution to agriculture. The link between ozone air pollution and the productivity of commercially grown crops in managed ecosystems (farms) has been well established by a series of field experiments known as the National Crop Loss Assessment Network (Heck et al. 1983). Several authors have combined the ozone concentration/yield loss relationships from the NCLAN experiments with economic models of the agricultural sector to estimate the value of reducing ozone pollution (Adams, Hamilton, and McCarl 1984, Adams and McCarl 1985; Kopp, et al. 1985)

In each of these studies, the predicted effects of changes in ozone levels on crop yields were used as inputs in an economic model of the supply and demand for agricultural products. A reduction in ozone pollution results in an outward shift of the supply curves of the affected crops. Changes in supply curves lead to changes in outputs, prices, consumers' surplus and producers' surplus.[8] The economic value of the change in ozone levels is the sum of the changes in producers' and consumers' surpluses. While these studies are not valuing any specific ecosystem function, they are valuing the effect of changes in ozone pollution on the set of ecosystem functions that determine the size of the service flows from this managed ecosystem to people.

Recreational Fishing

The third approach involves estimating changes in the demand for and the value of access to recreational fishing as a consequence of changes in fishing success or catch rate. A number of studies of the demand for recreation activities, based either on some form of travel cost model or contingent valuation questions, have investigated how revealed or stated values for recreational fishing change with observed or postulated changes in catch rate.[9] These studies show that it is possible to estimate the values of changes in nonmarket ecosystem service flows that people use by visiting the ecosystem. What is missing in all of these studies is the link between ecosystem functions and success at catching fish.[10]

Valuing a Species

Finally, a number of studies have employed contingent valuation questions to estimate the values that people place on the preservation or existence of various types of natural assets even though they may never visit or use the service flows from these assets. Such values are termed existence values or nonuse values. Nonuse values have been estimated for individual species such as grizzly bears and bighorn sheep (Brookshire, Eubanks, and Randall 1983), various marine mammals (Samples, Dixon, and Gowen 1986; Hageman, 1986), bald eagles (Boyle and Bishop 1987) and whooping cranes (Bowker and Stoll 1988).[11] Although the viability of populations of these species depends on the characteristics of the ecological systems of which they are a part, the values of these species have not been linked to any specific ecosystem functions.

In summary, economic valuation methods can be used to estimate people's willingness to pay for a variety of service flows from ecosystems, and substantial literature reports applications of these methods. In some cases, these values are linked to environmental quality variables that affect the underlying ecosystem functions (for example, ozone in agriculture). However, I am not aware of any studies that successfully make the link between the values of service flows and the ecosystem functions that support them.

Making the Link

Biological and ecological scientists may be better able than I to suggest ways to make the connections between the service flows described above and the underlying ecosystem functions that support them. Looking back on the discussions and presentations at the workshop, I summarize several possible connections.

One of the ecosystem functions discussed during the workshop is the denitrification performed in aquatic ecosystems. It was suggested that a decrease in this function could result in an increase in nitrate concentrations and a decline of submerged aquatic vegetation. This loss might lead to impacts on commercial and recreational fisheries. Since fish harvest is a service flow, the relationship between nitrate concentrations and fish populations could provide a basis for valuing the denitrification function in terms of its contribution to the economic value of fish harvest. If denitrification also supports other economically valuable service flows, the approach described here provides only a partial measure of the value of this ecosystem function. The total value of the denitrification function would be the sum of its contributions to the values of all of the affected service flows.

Another example comes from the ongoing Patuxent Estuary Project described by Geoghegan at the workshop. One part of this project involves the development of an economic model to predict changes in land use as a function of economic variables and land use regulations. Changes in land use will be used as inputs into an ecosystem model of the hydrological cycle and nutrient fluxes. This model will predict changes in a number of ecological variables. Finally, there will be an effort to estimate the economic values of the resulting ecological changes.

One set of outputs from the model will be changes in several water quality variables. The researchers hope to use these results to predict changes in fish abundance for species of recreational and commercial significance. Existing economic methods can be used to value the changes in the service flows provided by these fish. Changes in land use and the associated ecological changes will also change the terrestrial landscape. The researchers want to estimate the value of these landscape changes. However, one problem they face is that the ecosystem model they are using aggregates all terrestrial plant production into total biomass in kilograms. In this respect, the ecological model does not provide enough detail to take account of the differences in the aesthetic values that people place on different plant species and plant communities.

Another ecosystem function discussed during the workshop is the role of top carnivores in controlling populations of large herbivores such as deer, moose, and elk. It was suggested that the economic cost of the loss of this function in the northeastern U.S. could be found by estimating the economic damages caused by the excess population of herbivores. These damages include the losses of commercial and ornamental trees and plants due to browsing and the costs associated with animal-vehicle collisions.

But there is another economic dimension to the loss of this ecosystem function. Deer and moose herds also support other economically valuable service flows: in particular, recreational hunting and wildlife observation. There is evidence that these are highly valued service flows. Thus, it is not at all clear that the loss of top carnivores in the northeast has a negative economic value, all things considered.

Another example of a potential linkage involving the loss of predation comes from a recent paper by Holling (1994). He suggests that a decline in predation of spruce budworm larvae by songbirds might be contributing to the increase in damages to the commercial spruce-fir forests of eastern Canada from defoliation. If the link between song bird populations and spruce budworm outbreaks could be established, the economic value of song bird predation could then be estimated by looking at changes in forest harvests, prices, and consumers and producers surplus. The decline in song bird populations appears to be related to fragmentation of forest habitats in North America (Askins 1995; Robinson, et al. 1995). If that is the case, then the links described here could provide the basis for estimating one component of the economic value of unfragmented forest habitat.

I hope that these examples have demonstrated the potential for ecologists and economists to work together to investigate the links between ecosystem functions and the ecosystem services that people value. I suggest this as a potentially fruitful area for collaborative research. It will require that the economists develop a better understanding of ecological concepts and that the ecologists accept both the strengths and limitations of the economic valuation paradigm.

References

Adams, R. M., S. A. Hamilton, and Bruce McCarl. 1994. *The Economic Effects of Ozone on Agriculture.* Environmental Protection Agency, Corvallis, OR.

Adams, R. M., and B. A. McCarl. 1985. Assessing the benefits of alternative ozone standards on agriculture: The role of response information. *Journal of Environmental Economics and Management* 12(3):264–276.

Askins, R. A. 1995. Hostile landscapes and the decline of migratory songbirds. *Science* 267:1956–1957.

Barbier, E. B. 1994. Valuing environmental functions. *Land Economics* 70(2):55–173.

Bell, F. W. 1989. *Application of Wetland Evaluation Theory to Florida Fisheries.* The Florida Sea Grant College, Tallahassee, FL.

Bergstrom, J. C., et al. 1990. Economic value of wetlands-based recreation. *Ecological Economics* 2(2):129–147.

Bowker, J. M., and J. R. Stoll. 1988. Use of dichotomous choice, non-market methods to value the whooping crane resource. *American Journal of Agricultural Economics* 70(2):372–381.

Boyle, K. J., and R. C. Bishop. 1987. Valuing wildlife in benefit-cost analysis: A case study involving endangered species. *Water Resources Research* 23(5):943–950.

Braden, J. B., and C. D. Kolstad (eds.). 1991. *Measuring the Demand for Environmental Quality.* Elsevier, Amsterdam, The Netherlands.

Brookshire, D. S., L. S. Eubanks, and Alan Randall. 1983. Estimating option price and existence values for wildlife resources. *Land Economics* 59(1):1–15.

Crocker, T. D., and J. Tschirhart. 1992. Ecosystems, externalities, and economics, *Environmental and Resource Economics* 2(4):551–567.

Fisher, A. C., and W. M. Hanemann. 1986. Option value and the extinction of species. In V. Kerry Smith (ed.), *Advances in Applied Micro-Economics*, Vol. 4. JAI Press, Greenwich, CT.

Freeman, A. M. III. 1993a. *The Economics of Valuing Marine Recreation: A Review of the Empirical Evidence.* Bowdoin Economics Working Paper 93–102.

——— . 1993b. *The Measurement of Environmental and Resource Values: Theory and Methods.* Resources for the Future, Washington, DC.

——— . 1995a. The economic valuation of coastal resources supporting recreation. In C. S. Colgan (ed.). *Sustaining Coastal Resources: The Roles of Economics and Environmental Sciences.* Edmund S. Muskie Institute of Public Affairs. Portland. ME.

——— . 1995b. The benefits of water quality improvements for marine recreation: a review of the empirical evidence. *Marine Resource Economics* 10(4):385–406.

——— . III. R. H. Haveman. and A. V. Kneese. 1973. *The Economics of Environmental Policy.* John Wiley. New York. NY.

Gottfried. R. R. 1992. The value of a watershed as a series of linked multiproduct assets. *Ecological Economics* 5(2):145–161.

Hageman. R. K. 1986. Economic Valuation of Marine Wildlife: Does Existence Value Exist? Paper presented at the Association of Environmental and Resource Economists Workshop on Marine Pollution and Environmental Damage Assessment. Narragansett. RI. June 5–6.

Heck. W. W.. et al. 1983. A reassessment of crop loss from ozone. *Environmental Science and Technology* 17:572a–581a.

Holling. C. S. 1994. An ecologist's view of Malthusian conflict. In K. Lindahl-Kiessling and H. Landberg (eds.). *Population. Economic Development, and the Environment.* Oxford University Press, New York. NY.

Kopp. R. J.. et al. 1985. Implications of environmental policy for U.S. agriculture: The case of ambient ozone standards. *Journal of Environmental Management* 20(4):321–331.

Kopp. R. J.. and V. Kerry Smith (eds.). 1993. *Valuing Natural Assets: The Economics of Natural Resource Damage Assessment.* Resources for the Future. Washington. DC.

Lynne. G. D.. Patricia Conroy. and Frederick J. Prochaska. 1981. Economic valuation of marsh areas for marine production processes. *Journal of Environmental Economics and Management* 8(2):175–186.

McConnell. K. E. 1993. Indirect methods for assessing natural resource damages under CERCLA. In Raymond J. Kopp and V. Kerry Smith (eds.). *Valuing Natural Assets: The Economics of Natural Resource Damage Assessment.* Resources for the Future. Washington. DC.

Robinson. S. K.. et al. 1995 Regional forest fragmentation and the nesting success of migratory birds. *Science* 267:1987–1990.

Samples. K. C.. John A. Dixon. and Marsha M. Gowen. 1986. Information disclosure and endangered species valuation. *Land Economics* 62(3):306–312.

Schulze, W. D. 1993. Use of direct methods for valuing natural resource damages. In Raymond J. Kopp and V. Kerry Smith (eds.), *Valuing Natural Assets: The Economics of Natural Resource Damage Assessment.* Resources for the Future, Washington, DC.

Simpson, R. D., R. A. Sedjo, and J. W. Reid. 1996. Valuing Biodiversity for Use in Pharmaceutical Research. *Journal of Political Economy* 104(1):163–185.

Smith, V. K. 1988. Resource evaluation at the crossroads. *Resources* 90:2–6.

Notes

1.This and the next definition come from Webster's New World Dictionary 1988. Third College Edition Simon and Schuster, New York, NY.

2.   The development of short stemmed varieties of rice has the effect of increasing the proportion of the photosynthesis carried out by each plant that is useful in this sense.

3.   For an elaboration of this idea, see Crocker and Tschirhart (1992) and Barbier (1994).

4.   See Toman (this volume) for a further explanation of this framework. Toman also discusses some of the underlying issues concerning valuation methods and the use of economic values to guide policy choices, for example, the question of the distribution of benefits and costs, discounting of future benefits and costs, and intergenerational equity.

5.   In addition to the types of service flows described above, ecosystems serve as stores of potentially valuable information in the form of genetic data. Economists have been turning their attention to the question of the economic value of this information. For two interesting examples of these efforts, see Fisher and Hanemann (1986) and Simpson, Sedjo, and Reid (1996).

6.   For detailed expositions of the principal methods for valuing environmental changes in general, see Braden and Kolstad (1991) and Freeman (1993b). McConnell (1993) and Schulze (1993) also describe these methods in the specific context of assessing the damages to natural resources.

7.   The net monetary value is the market price minus the cost of labor, capital, etc. used in harvesting the fish.

8.   Consumers' surplus is the difference between the total amount they would have been willing to pay for a specified quantity and their actual expenditure. Producers' surplus is the difference between total revenue and the variable costs of production.

9.   For a review of studies dealing with marine recreational fishing, see Freeman (1993a) and Freeman (1995a).

10.   Some of the studies reviewed in Freeman (1993a) attempt to introduce various measures of water quality, but these efforts are not generally successful.

11.   There is also a contingent valuation study of the value of a set of recreation services associated with the coastal wetlands of Louisiana. See Bergstrom et al. (1990).

CHAPTER TWELVE

Ecology and Public Policy

Michael K. Orbach

Introduction

This paper grew out of the session entitled, "How Ecological Concerns are Translated into Social Decision Making" at the symposium which generated this volume. The general question posed in the session was, "Why perceptions of economic inefficiency and ecological risk are not translated into appropriate policy action." In addressing this question, the general response of the session members was this: "Appropriate policy action" is generated through systems of governance, not systems of science alone, and those systems of governance are primarily based on human values and human ecology. In this sense, environmental policy is *social* policy (Maiolo and Orbach 1982).

The concern in this paper is with the relationship between science and scientists and the process of decision making in public policy with respect to environmental and natural resource issues, and in particular the role of the social sciences in incorporating human values into that process. One aspect of this subject is the question of "value," which in the recent economics literature has been broadly construed to include measures of perceptions of relative benefit derived and costs incurred by humans from various uses and states of an environment (see Toman, and Geoghegan and Bockstael, this volume). In addition, there is the question of whether "value" exists apart from any relationship of thought, perception and use to human beings (Norton 1982). Finally, there is the anthropological sense of the term "value," which refers to a human cultural rule or standard. This last, more general sense of the term often more closely reflects the way humans set the cultural rules—the "policies"—which guide their behavior. The concept of "value," with any of these connotations, is central to the way we develop and implement our environmental policies.

In this paper we will place the issue of human decision making with respect to our attitudes towards and use of our environment, or ecosystems, into the broad framework of public policy making. We will first define the terms "policy," "value," "ecology," and "ecosystem." Then we will outline the nature of the public policy process with respect to environmental issues and problems. Third, we will discuss the role of science and scientists, both natural and social, in this process. Finally, we will comment on the "reconciliation" of natural and social science and public policy with respect to environmental issues.

Human Decisions, Policy Making, and Ecosystems

The management of human behavior within ecosystems which contain both human and nonhuman components occurs through the mechanisms of both private and public policy making. We use the term "policy" to mean a rule of human behavior based on cultural perception and preference. Private sector policies are those rules that we use to guide our behavior concerning private property and behavior not subject to the "public trust." Public sector policies are those rules which guide the behavior of all members of our society concerning "public goods" (i.e. common property resources) and "the public good" (human health and safety, sustainable environments, etc.), and which are developed in the context of organizations formed to attend to the 'public trust'—the government (Slade et al. 1990). In either case, what is being "managed" by the policy are not the nonhuman elements of the ecosystem directly, but the human elements and their behaviors associated with that ecosystem.

When we refer to the "environment," or "ecosystem," we are generally adopting Odum's (1971) definition: "Any unit that includes all of the organisms (ie., the biotic and abiotic 'communi-

ty') in a given area interacting with the physical environment
. . . within the system is an ecological system or ecosystem."
For our purposes, we will explicitly distinguish between the human and nonhuman elements of the ecosystem, and lump the
nonhuman elements together under the term "physical ecology."
We thus use the terms "physical environment" and "physical
ecology" to refer to the nonhuman components and processes in
an environment, including the chemical, physical, and biological
resources of that environment. We will use the term "human
ecology" to refer to the human components and processes in an
environment—the people and their beliefs, perceptions and behaviors. Our focus is an anthropocentric one, following the approach of the (human) cultural ecologists (Harris 1968; Steward
1972; Vayda 1969; White 1949).

These two "systems"—the human and the nonhuman—are of
course systemically interrelated with respect to natural resource
policy issues (Miller et al. 1987). We make the analytical distinction here because all policy making is anthropogenic—it is generated by humans—and the only direct effect of policy is to
change human behavior. Thus, in our discussion of policy making we wish to analytically isolate the objects of that process
within the ecologic system. The term "environmental, or ecosystem, policy" will refer to human decision making with respect to
human perception or use of the physical ecology of an environment or ecosystem.

All ecosystem policy and management, whether emanating
from the private or public sector, necessarily involve human cultural values, and decision making based on those values (Cubbage and Brooks 1991). The decisions which comprise those
policies necessarily reflect underlying human values. As we noted in the introduction, however, the term "value" has several different connotations. An anthropologist would define "value" as a
culturally defined cognitive (how we think) or behavioral (how we
behave) standard. This is very different from the sense in which
an economist might use the term "value," which is most often
expressed as a price, and has the connotation of the marginal
change in costs or benefits associated with a particular human
behavior. The fact that humans wish to conserve natural resources, to ensure their continued availability and "health" into
the future, is a cultural rule defined by humans themselves in
the anthropological sense. No matter how much we might think
that this is a "natural" or "logical" rule, it is in the end a human

perception and evaluation subject to many different forms of human interpretation and implementation. Rules which allocate the use of natural resources among different human uses and users, on the other hand, are more often based on decisions concerning the marginal cost or benefit of that use in the economical sense. Some of these rules may be based on judgment of some measurable remunerative stream, such as whether forestry or fishery products will yield more dollars in one form of harvest or another.

Policies—rules—which assign different categorical importance to different components of the physical environment, such as those in the Marine Mammal Protection Act (MMPA), are based on the cultural values assigned by humans to particular components of the physical environment. In the case of the MMPA, an absolute preservation "value" for marine mammals, irrespective of any notion of remunerative "value," was also largely independent of the biological status of the species involved. That is, marine mammals were to be completely protected even if they were not threatened or endangered. The Endangered Species Act, of course, expressed our society's cultural preferences with regard to human behavior causing the endangerment of species, a preference also generally not tied closely with remunerative issues.

On the other hand, policies which create a system to make decisions concerning both the consumptive and non-consumptive uses of certain components of the physical environment, such as the Magnuson Fisheries Conservation and Management Act, the Multiple Use and Sustained Yield Act, the National Forest Management Act, and the Federal Land Policy and Management Act are more often based on the marginal remunerative costs and benefits accruing from different human behaviors towards those resources.

Human Ecology, Ecosystem, and Environment

As defined above, we are using the term "human ecology" somewhat differently from the classical literature on human, or cultural ecology (Vayda 1969). This literature most often uses these terms to define the total set of relationships among humans and what we are referring to here as the physical ecology of an environment, or ecosystem. We are using the term "human ecology" more restrictively to mean only the humans and their behaviors for two reasons. The first is to emphasize the point that public policy directly governs only human behavior, and only indirectly the states or behaviors of the nonhuman elements of an environment or ecosystem. The second is to maintain the conceptual and analytical distinction between the structures and processes of human behavior as distinct from the structures and process of nonhuman elements of an environment or ecosystem.

We have further defined the term "environment," or "ecosystem" to mean any circumscribed physical ecology (nonhuman components and processes) and the human beings and their behaviors (human ecology) associated with that physical ecology. To be entirely conceptually and analytically complete we could also use the mirror image of this definition: that "environment" or "ecosystem" may mean any circumscribed human ecology (human components and processes) and the physical ecology (nonhuman components and processes) associated with that human ecology. The latter is in fact the perspective often taken by the social and policy sciences in their discussion of environmental issues. Finally, we will use the terms "environment" and "ecosystem" to be synonymous with one another, but for convenience will use the latter, "ecosystem" for the remainder of this paper.

Just as a forest or fisheries ecologist views all of the physical, chemical, and biological components and processes of a forest or fishery population and habitat as interrelated, so must we view the human components and processes associated with such populations or habitats as being interrelated. Within any given ecosystem encompassing both human and nonhuman components, in order to study policy making we must focus on the human components and processes. When we speak of "valuation" or "governance" (see Hennessey, this volume), it is human belief, perceptions and behaviors of which we are speaking.

The human ecology of any ecosystem has two broad subcomponents: 1) the human constituencies of the physical ecology of that ecosystem, for example the people who live in, use, or otherwise are concerned directly or indirectly in their beliefs and behaviors with that physical ecology, and 2) the humans who make up the organizations which develop and implement the policies intended to govern the behavior of the constituencies defined in 1). The human ecology of an ecosystem, in skeletal form, is represented in Figure 12–1. For the purposes of this paper we

[Physical Ecology]

Constituent Component

   Direct

      Industries within the Physical Ecology

      Citizens within the Physical Ecology   **Scientific Component**

      Interest Groups (Internal)   **Natural Scientists**

   Indirect   **Social Scientists**

      Industries External to the Physical Ecology   **Environmental**

      Citizens External to the Physical Ecology   **Engineers**

      Interest Groups (External)

Governance Component

      International

      Federal

      State

      Local

      Private Sector

Figure 12–1.   The Human Ecology of an Ecosystem [Physical Ecology]

have separated "scientists" as a distinct component of this human ecology, a component related to both the constituent and governance components.

This human ecology includes people who may be very remote from the physical ecology in question. Even though it is the people resident or active in the physical ecology itself whose behavior is most directly affected by policy decisions concerning that physical ecology, those policies may also affect—and be affected by—people who never physically encounter that physical ecology. Policies affecting coastal and marine areas, for example, affect people distant from those areas because they *care about*— that is, because they have some value orientation towards—the resources of those ecosystems, and because they may be participants in the policy-making process even though they never personally interact with those resources.

Offshore oil and gas policy is a good example of such a policy arena. Offshore oil and gas policy affects people whether they live at the coast or not, and noncoastal residents certainly play a part in the formation of policies concerning offshore oil and gas; witness the case of oil spills. On the other side, public policy makers whose actions affect coastal constituencies are often themselves very remote from coastal and marine areas. This fact has significant implications for the ability of those policy makers to adequately understand coastal constituencies and to formulate, implement, and evaluate policy. These same comments apply to every policy arena—fisheries, forestry, water quality, hazardous waste, and so on. This phenomenon is also a part of the mismatch between an environmental problem or issue and the governance systems which must make policy concerning that problem or issue. Our governance system boundaries often do not match our ecosystem boundaries in either their physical or human dimensions, a point to which we will return below.

To understand the human component of any ecosystem, and to understand the policy and management—the governance—process itself with respect to that ecosystem, we must understand the beliefs and behaviors of all of the people in the human ecology outlined in Figure 12-1. This is an error both of commission and omission that we have made in environmental studies in general, and in the formulation of environmental policy in particular (Millsap 1984; NRC 1989). We have been severely imbalanced in our collection and analysis of data and in our view of the public policy process in favor of natural science. The underrepresented understanding of human ecology is the province of social science and the humanities. Humans and their beliefs and behaviors form the human ecology of any given ecosystem, and are the major factors in the process of governance of the human behavior with respect to that ecosystem, which is our subject here.

The Public Policy, or Governance Process

The human decision-making process which constitutes public policy, or governance, has certain characteristics. It is continuous, resolving rather than solving problems and issues. It is complex and difficult to bound analytically. It is purposive in that it is constructed to alter human behavior. If everyone were behaving voluntarily according to the common values of the larger society, there would be no need for public policy or governance (Anderson 1984). Such alteration can result either in changes in

human belief and perception or in changes in human behavior itself. Belief, perception, or behavior can be altered through incentive or education, in the broad senses of those terms. But only behavior, not thought or perception, can be directly altered through
regulatory sanctions. Any governance system is based upon beliefs, perceptions, and behaviors, and it is often as important to
change human belief and perception through education and debate as it is to change human behavior through regulation.

The decision-making process through which we make public
policy in the United States is called representative democracy,
and has a number of structural characteristics (bi-cameral leg-
islature; executive agencies; courts; confederation of states; constitutional government). The consideration of environmental
problems or issues occurs in the context of this system of governance. The decisions made within this system are subject to cultural and societal context, legal and administrative configuration, and—to be emphasized in particular here—the nature of
the scientific input to the governance process. In this sense
these decisions are unavoidably "political," with a non-perjora-
tive connotation of that word. When an individual or group ex-
presses the desire to get the politics out of environmental decision making, they are making either an uneducated or
unrealistic demand. They may *mean* that they wish more accurate data or information were available to the process, that the
process should be more "rational" in the sense that political scientists or economists use that term (Lindbloom 1959). They may
mean that they disagree with the particular set of value judgments that occurred in a particular case of environmental decision making, or that a particular governance process has not occurred in an ethical manner. Conceptually and practically,
however, one cannot "get the politics out" of environmental decision making—it is an inherently political governance process.

There is also the general issue of boundary definition in systems of governance. In natural resource management we most
often bound our discussions by considering all of the human behaviors associated with a particular wildlife population or habitat, or with a political or administrative boundary such as a city,
county, park or forest. These political and administrative boundaries may be related to physical ecosystemic properties, but
most often are not. It is much less often that we bound our discussion by considering all of the populations and habitats associated with a particular groups of humans. So, for example, the
patterns of behavior of migratory salmon or king mackerel pop-

ulations and the humans who exploit them at some point in their migration, and at some physical location within their habitat, provide the boundary definitions for traditional ecosystem management. Our federal fisheries laws instruct us that, " . . . an individual stock of fish shall be managed as a unit throughout its range . . . " (MFCMA Sec. 301(3)). This is as opposed to bounding our management by the patterns of behavior of commercial or recreational fishermen who "migrate" to fish for several different fish populations in locations within the habitats of those species. Our natural resource management laws do not instruct us to treat human populations as a unit of management, although we sometimes gravitate in that direction for practical purposes. When we do achieve ecosystem management, however, we more often manage, as a group, all of the human behaviors associated with a particular physical ecosystem, not behaviors associated with all ecosystems used by a particular group of people. Thus, even though policy making is anthropocentric in principle, it is not in practice, at least in the sense of boundary definitions.

Rivers, fish and wildlife, airsheds, forests, and virtually every other physical ecologic systems cross municipal, state, and national boundaries. Various attempts have been made to address physical ecologic boundaries (the National Estuary Program; River Basin Commissions), particular resource systems (Regional Fishery Management Councils) or administrative configurations (the National Biological Survey) which cross political or administrative jurisdictions. These attempts reflect the adaptive nature of human ecologic systems, analogous to the adaptive nature of physical ecologic systems; both are complex and continuous. At the moment we must deal with local (municipal, county), state or provincial, tribal and national boundaries and the governance systems which they define. Regional or ecosystem-defined environmental issues share the same status as international "law": both can be conceptualized on a system-wide basis, but any regulations must be ultimately codified within an existing political jurisdiction to be binding on human behavior (Hennessey, this volume).

The Role of the Constituencies

We define a constituent of a policy process as any person or group of people in a society who has access to that process, or who is affected by the policy decisions of that process. "The public" is the most broadly defined constituency of public policy, and the definition narrows as the group or category of constituency shrinks. Constituencies have multiple points of access to the governance system, to the process of deciding which *values*—in both the cultural and economic senses—upon which to base the rules of governance, and to supply data and information to the process in deciding how to apply those rules to particular situations of governance (Burch 1976). Let us emphasize that environmental policy and management is a *governance*, not a *scientific* process. Constituents can vote; they can join political parties, environmental organizations and other interest groups; they can attend public hearings and scoping meetings; they can contact legislators and administrative agency personnel directly; they can work within local communities to change values, beliefs, perceptions and behaviors; and they can bring suit through the court system (Wengert 1976). All of these are *legitimate* and, one could even argue, *necessary*, roles of constituencies.

One common problem with governance systems in general, and ecosystem policy and management in particular, is that in the past these systems have not been structured to include the constituencies in the governance process from top to bottom, beginning to end. Constituent input is critical in identifying the values which should underly the decision-making process. Constituents provide data and information to the process, although that data may not be deemed "scientific," and they can also be involved in the process of setting policy for the funding and conduct of science (Peterson 1984). Constituents may also be involved in the governance system itself, through the mechanism of "co-management" (Pinkerton 1989). It is possible to view constituencies as "clients" of professionals in the policy-making system, about whom some piece of data or information is necessary, or whose perceptions and interests must be included in the policy process.

What often happens in our policy-making process is that the process is captured by elected officials, scientists, administrators, advocates, or technicians who view environmental policy making either as a straightforward technical problem or as a set of value judgments whose configuration is known, and therefore need not have the benefit of input or collaboration from "the public" or from a full range of constituencies. This is almost always a fatal error. Even if the policy is made and no obvious deleterious effects on the physical ecology or constituencies occur from that particular policy in the short term, the fact that the policy makers have not, as a matter of principle, been sensitive to the need and right of the public and the full range of constituencies to be involved in and informed of the process will eventually lead to legitimate protests against the policy-making system.

On the other hand, care must be taken as to how constituencies are represented in the public policy process. The "public trust" cannot be violated by yielding undue influence over the governance process to particular constituencies, be they individuals, corporations, or environmental groups, influence which often bears such labels as "nepotism" or "conflict of interest." The National Environmental Policy Act (NEPA), the administrative procedures acts, both federal and state, and the majority of our post-1960s environmental legislation have made provision for constituent, or "public" input into administrative rule making in ways which are intended to both preserve the public trust and provide substantive participation by the public. Such provisions, however, have not been completely or uniformly applied, and in their current form may ultimately be insufficiently structured to allow for a full range of constituent input to policy making (Sabatier 1988).

The Role of Science

Similarly, NEPA and the majority of the post-1960s environmental legislation have specific provisions for scientific input, generally admonishing for the use of the "best scientific information available." This has not always come to fruition, for myriad reasons which are beyond the scope of this paper. However, discussion of two aspects of the role of science in policy making, and in governance in general, are in order here.

First, if such scientific input is to be used effectively it must be the truly interdisciplinary science referred to in NEPA. In particular, the question of value, belief, perception and behavior—the human ecology—that is central to all governance and to our discussion of environmental policy and management here can *only* be answered with thorough, high-quality social science work (Millsap 1984; Weiss 1977). Physical and natural scientific input, which contributes to our knowledge of what we have termed in this paper the "physical ecology" of an ecosystem, is of course necessary to rational policy and management. However, it is the historic lack of input from the social sciences concerning the human ecology of ecosystems and governance which leads to conferences such as that which generated this volume. Many factors have led to this situation, including the nature of the "subcultures" within the scientific community itself which often do not communicate or agree on basic definitions and principles of the conduct and application of the scientific enterprise (Fortmann 1990; Caldwell 1990). We do not want to belabor this point because the symposium which led to this volume is in fact an example of an attempt to approach this challenge, by including a wide range of both natural and social science disciplines. However, much progress is still to be made in this area, in environmental policy making in all of its aspects. For example, the funding for social science research to supply data and information to the policy process lags far behind that of the natural sciences, and the balance of professional staff in natural resource management agencies and appointments to scientific advisory committees is tipped drastically in favor of the natural sciences.

It is instructive to view the original title of the symposium upon which this volume is based, "Reconciling Economics and Ecology," and the hope that such a reconciliation might help solve the dilemma of "scientific" policy making, in this light. In this author's view, what occurred at the symposium was the realization that the inclusion of the economic perspective was one more significant step in the rational conceptualization of, and the inclusion of science in, environmental policy making, but far short of the number of steps necessary. In particular, recent advances in the concept of "valuation" were found to be useful in ordering the economic value of benefits humans obtain from the physical ecology of an environment, but less useful in fully constructing the policy-making process from the cultural values humans hold for that physical ecology or for the policy-making process itself. Significant progress was made in the specification of what data and conceptual and analytical contributions could be made by economics, and what data and conceptual and analytical contributions might have to be made by the other social sciences.

The second point is that no science—physical, natural, or social—will yield the "answer" to any governance question, including those involving environmental policy. We often look to scientists for the "answer" to environmental policy questions. The problem is that "answers" to governance questions must come through the governance process, which takes individual items of scientific data and information and uses them as pieces of a larger puzzle involving human values and human decision making. A scientist might be able to determine how many fish of a particular species are in the water, but not necessarily how many fish of that species *should* be in the water. A scientist might be able to tell us how our wetlands are changing, but not necessarily how they *should* change. A scientist might be able to tell us how users of a forest or water source might be affected by changes in that forest or water source, but not how they *should* be affected. In general, scientists can tell us the characteristics of things and the potential effects of different human behaviors on those things; they cannot tell us which configuration of those characteristics is *best* in human value terms, or which effects are *better* or *worse* in human value terms (Caldwell 1990; Hammond and Adelman 1976; Shrader-Frechette and McCoy 1994). In addition, in the past scientists have not always been careful to distinguish between their own personal value systems and those of other constituents in the larger society; that is, they have failed to distinguish between their role as scientist and their role as advocate. Social scientists can, in fact, through their research yield insight into the characteristics and distribution of values, both cultural (anthropologists, sociologists) and economic (economists), and of the policy-making process itself (political science, public administration). The folding of those values into the policy-making process, however, is not the formal function of science. That is the province of the governance process, of the integration of scientific information with human value-based decision making.

The "Reconciliation" of Natural and Social Science and Public Policy

In light of the fact that environmental policy making is a human value-based decision process, and that this process involves intimate knowledge and documentation of both the physical and the human ecologies of an ecosystem and their relationship to one another, what are the principles upon which an approach to environmental policy making with effective scientific input should be based?

1. The Total Ecosystems Perspective—The process of environmental policy making should be "integrated" across physical and human ecologic boundaries; across sectors (fisheries, oil and gas, forestry, etc.); across environmental engineering and scientific disciplines (natural and social scientific); across lines of governance (local, state, national) and across tasks (legislation, administration, research, monitoring, public involvement) (Cicin-Sain 1993). As we discovered in the conference which led to this volume, no single disciplinary perspective (i.e., economics and "valuation") will provide us with "the answer." The complimentary contributions across these sectors can provide a sound point of departure for more effective public policy deliberations. This argues for the creation of new partnerships among members of the policy and management, science and engineering, and private sector communities.

2. The Democratic Nature of Our Governance System—For scientists to participate effectively in the governance process, they must recognize their potential for different roles in that process (Jasanoff 1990). If they act as scientists, their role is largely confined to providing specific pieces of data and information which will then be used by others in value-based decision making. If they act as policy makers or administrators, they can have a role in those subsequent portions of the decision-making process. And, of course, scientists can always act as advocates in their role as private citizens or on behalf of a particular community or organization, in which case they will be expressing *values, beliefs, and perceptions* on the basis of which the 'scientific' information might be *interpreted*. It is necessary for the scientists to clearly recognize which role they are assuming, and to equally clearly communicate the nature of that role to others. If these roles are not clearly distinguished, the science itself may be called into question on the basis of inherent bias. Ultimately, however, scientists will have to recognize that they are not the final decision makers, but rather a specific portion of the total governance process.

Moving beyond the role of science to the role of scientists as client-servers for constituencies of the governance process, scientists must consciously encourage the inclusion of those constituencies in the policy process. This is a delicate question, because it is rife with possibilities for bias and ill-conceived advocacy. However, at every step in the environmental policy process scientists, both social and natural, should encourage review and participation by "the public" in the process. This includes everything from how science is funded, to how the governance process is constructed, to the consideration of specific environmental issues.

3. The "Translation" Function—The process of taking scientific data and information and formatting and interpreting it for the policy-making process is an area of expertise in itself. Not every scientist is skilled in the application of their scientific results, just as the "experts" in the application process may not be experts in the process of doing science. Each participant in the process must recognize the need for these different skills and roles, including the role of environmental engineers, and respect the role of the others even though the particular outcome, or 'answer' from a policy process might not fit the value-orientation of any particular participant. This is the difference between saying, "They didn't *understand* my data," and saying, "They *interpreted by data through a different value system.*"

Conclusion

This paper began by pointing out that environmental policy, like all other public policy, is *social* policy in that the process of governance takes scientific and other data and combines them in a human value-based, democratic process the ultimate purpose of which is to alter human behavior. This argues for an increased emphasis on the notion of "value" in all of its connotations, some of which are economic and generally remunerative and others of which are cultural and generally non-remunerative. In addition, to understand an ecosystem or environment, and to successfully apply the notion of "value," we must understand and document both the human behaviors and the nonhuman processes associated with that environment. This argues for the separation of the human and nonhuman ecology of an ecosystem for analytical purposes, and for an emphasis on the fact that it is only the human behaviors that are directly affected by policy making.

Finally, all of the professionals and citizens involved or concerned with an ecosystem must be involved in its governance, and that involvement implies a division of labor and collaboration among different scientists, between scientists and policy makers, and among scientists and policy makers and private-sector constituents. Each must respect the different roles of the others in this process to effectively combine science and human values in the environmental governance process.

References

Anderson, J. 1984. *Public Policy-Making*. Holt, Rinehart and Winston, New York, NY.

Burch, W. 1976. Who participates? *Natural Resources Journal* 16:41–54.

Caldwell, L. 1990. *Between Two Worlds: Science, the Environmental Movement, and Policy Choice*. Cambridge University Press, Cambridge, UK.

Cicin-Sain, B. (ed.). 1993. Integrated Coastal Management. Special issue of *Ocean and Coastal Management* 21:1–3.

Cubbage, F. and D. Brooks. 1991. Forest resource issues and policies: A framework for analysis. *Renewable Resources Journal* Winter: 17–25.

Fortmann, L. 1990. The role of professional norms and beliefs in the agency-client relations of natural science bureaucracies. *Natural Resources Journal* 30(3):361–380.

Hammond, K. and L. Adelman. 1976. Science, values and human judgment. *Science* 194:389–396.

Harris, M. 1968. *The Rise of Anthropological Theory: A History of the Theories of Culture*. Cromwell, New York, NY.

Jasanoff, S. 1990. *The Fifth Branch: Science Advisors as Policy-Makers*. Harvard University Press, Cambridge, MA.

Lindbloom, C. 1959. The science of muddling through. *Public Administration Review* 19:79–88.

Magnuson Fishery Conservation and Management Act, P.L. 94-265, as amended (1990).

Maiolo, J. and M. Orbach (eds.). 1982. *Modernization and Marine Fisheries Policy.* Ann Arbor Science Publishers, Ann Arbor, MI.

Millsap, W. 1984. *Applied Social Science for Environmental Planning.* Westview Press, Boulder, CO.

Miller, M. R. Gale and P. Brown. 1987. *Social Science in Natural Resource Management Systems.* Westview Press, Boulder, CO.

Norton, B. 1982. Environmental ethics and the rights of future generations. *Environmental Ethics* 4(2):319–330.

National Research Council. 1989. *The Adequacy of Environmental Information for Outer Continental Shelf Oil and Gas Decisions: Florida and California.* Report of the Committee to Review Outer Continental Shelf Environmental Studies Programs, Board on Environmental Studies and Toxicology, Commission on Physical Sciences, Mathematics and Resources, National Research Council. National Academy Press, Washington, DC.

Odum, E. 1971. *Fundamentals of Ecology.* W.B. Saunders, Philadelphia, PA.

Peterson, J. 1984. *Citizen Participation in Science Policy.* University of Massachusetts Press, Amherst, MA.

Pinkerton, E. (ed.). 1989. *Co-operative Management of Local Fisheries: New Directions for Improved Management and Community Development.* University of British Columbia Press, Vancouver, BC.

Sabatier, P. 1988. An advocacy coalition framework or policy change and role of policy-oriented learning therein. *Policy Sciences* 21:129–168.

Schrader-Frechette, K. and E. McCoy. 1994. *Method in Ecology: Strategies for Conservation.* Cambridge University Press, Cambridge, UK.

Slade, D. (ed.). 1990. *Putting the Public Trust Doctrine to Work: The Application of the Public Trust Doctrine to the Management of Lands, Waters and Living Resources of the Coastal States.* Connecticut Department of Environmental Protection, New Haven, CT.

Steward, J. 1972. *Theory of Culture Change: The Methodology of Multilinear Evolution.* University of Illinois Press, Urbana, IL.

Vayda, A. 1969. *Environment and Cultural Behavior: Ecological Studies in Cultural Anthropology.* Natural History Press, Garden City, NY.

Weiss, C. 1977. *Using Social Science Research in Public Policy-Making.* Lexington Books, Lexington, MA.

Wengert, N. 1976. *Citizen participation: Practice in search of a theory.* Natural Resources Journal 16:23–40.

White, L. 1949. *The Science of Culture: A Study of Man and Behavior.* Farr and Strauss, New York, NY.

Epilogue

Norman L. Christensen, Jr.
R. David Simpson

A World of Change

Human beings have never had a greater impact on the natural world than they do now, and this impact will almost certainly grow. There may be disagreements between and among scientists and policy makers as to the extent and exact nature of human-wrought environmental changes, but few would deny that the human impact on the environment is an issue worthy of extensive study, debate, and, in many instances, public action.

Of course, it is true that humans have been modifying their environment for hundreds of thousands of years. And indeed, *Homo sapiens* is not the only species that changes the ecosystem it inhabits. One need only compare a free-flowing stream with a string of beaver ponds to be impressed with the impact a single species may have on its habitat. Our prehistoric ancestors also caused profound changes in their world. The extinction of megafauna in the Americas, Australia, the Pacific Islands, and other areas may have been largely the work of wanderers and voyagers who were also hunters. In more recent times, humans have hunted more species to extinction. We may have induced equal or greater losses in biological diversity inadvertently by displacing indigenous organisms with the flora and fauna we have husbanded for food or companionship. The shift to agrarian practices and fixed settlements resulted in other changes: the depletion of soils, deforestation, and overgrazing. The development of mining, metallurgy, dyeing, and primitive manufacturing introduced concentrated pollution.

While humans had induced great changes in their ecosystems before the industrial revolution, that era may seem in retrospect an age of innocence. Our ancestors committed some environmental blunders, but, by and large, their world was a pristine one. It was also one in which life was all too often "nasty, brutish, and short," however. Preindustrial humanity may have had little reason to be troubled by the "sustainability" of their ecosystems, but individual women and men had far greater reason to be concerned over the source of their next meal, the cause of a cough or sneeze, and the adequacy of their clothing and shelter.

One could argue endlessly as to whether the environmental problems of the present differ qualitatively, or only in degree, from those of the past. Two facts are self-evident, however. The first is that the spatial and temporal scales of modern environmental problems far exceed those of the preindustrial past. Consider chemical pollution. If it occurred at all in the past, it was confined to relatively small areas. The impacts of greenhouse gases and ozone-depleting chemicals about which we now worry, on the other hand, may truly be global. It is of course true that, as the bumper-sticker slogan has it, "extinction is forever," and our ancestors did wipe out some species. Modern technology raises the prospect of long-lived contamination that could not only wipe out some species but, conceivably, render large areas unfit for habitation by virtually all species. Perhaps, in the past, our forebearers could take comfort in thoughts such as "if we drive the game away here, there will be more nearby;" "after we have harvested this plot, we can let if lie fallow until it is restored;" or "poisoning this land or water will have little consequence, since we can plant our crops in the next river valley over." It is not quite so easy to be dismissive of modern environmental concerns. We may have serious reason to doubt that, if we eliminate still more species, the ecosystems comprised by surviving species will recover, or that, if we render one area permanently unusable with toxic or radioactive waste, other lands will be available to meet our needs.

The second incontrovertible fact about modern environmental problems is that they have evolved in tandem with the mastery of technology. Vast herds of American bison withstood millennia of Native American predation. A relative handful of nineteenth-century European-American hunters could not have come within an eyelash of eradicating them if they had not been armed with repeating rifles and whisked from killing field to killing field behind steam locomotives. The toll technological progress has exacted on natural ecosystems may be better illustrated by examples in which harm has been unintended rather than overt. The proximate cause of England's "killer fogs" was the coal burned in her "dark satanic mills." The rise of concentrated industrial production (and with it, concentrated industrial pollution) was itself compounded by innumerable factors, however. Improved transportation allowed both more distant distribution of industrial output and more widespread collection of industrial input; new sources of motive force freed factories from the necessity of location near fast-flowing water; improvements in agricultural production and distribution enabled large concentrations of workers to live at a greater remove from farms and fields. We could list many other components, but will summarize by saying that the characteristics of what is commonly identified as "economic progress" were also the preconditions for human-induced environmental damage on an unprecedented scale.

Technologies have given us capabilities to prolong and enrich human lives on one hand and to sometimes wantonly, more often negligently, endanger our ecological "life support system" on the other. We do not propose to argue that these technologies are good or bad. It is beyond dispute that, with the benefit of hindsight, we might have deployed many technologies differently, and perhaps some not at all. It is also indisputable that, for much of humanity, life expectancy and material comfort are higher in the twentieth century than in any other era of the past. It is impossible to compare overall well-being between different people and different times, but few of us would trade the wealth and relative security of our lives for those of our preindustrial counterparts. The issue at this late date is not whether the "natural" world is superior to the "artificial," but rather, what vestiges of the former we want or need to protect against the expansion of the latter. Depending on one's world view, she may rue the industrial growth that has trammeled nature or regard it as humanity's greatest achievement. Whichever the case, we cannot, as our forebearers might, claim ignorance of the environmental effects of our decisions.

While we are aware now that industrial expansion does have economic consequences, we still have far from perfect information about the extent of those consequences. We are aware that toxic pollution, nuclear radiation, habitat conversion, overexploitation, greenhouse and ozone-depleting gasses, and other products and by-products of economic activity can harm our environment. In some instances we have discovered quantifiable relationships between, for example, the concentration of certain toxins and their effects on animal and plant physiology. We are much farther from having any notion as to how much we must curtail economic activity in order to achieve any given improvement in environmental performance. In fact, we have little understanding of the long-term environmental consequences of the technological and environmental status quo.

Yet, we as a society (or as a group of societies) make decisions concerning environmental tradeoffs every day. As environmentalists remind us, these decisions are made at the real, if not often (or yet) realized, risk of apocalyptic consequences. From the other end of the political spectrum, we are reminded that excessive caution in the protection of environmental amenities has its consequences as well. Resources allocated to the protection of the environment may not be available for other purposes. Some of this forgone consumption may be frivolous, but some may be diverted from the basic needs of deserving individuals.

Ecologists and Economists

The papers presented here, as well as the workshop that produced them, provide ample evidence of the need for, as well as the challenges to, communication between economists and ecologists. Sustained provision of the goods and services derived from ecosystems depends not only on our understanding of the processes that underlie them, but on the existence of economic systems that provide appropriate incentives for the maintenance of sustainable ecological systems. Two challenges to collaboration between ecologists and economists are key.

First, there must be mutual understanding of the norms of scholarship and central paradigms in respective disciplines. For example, some ecologists impune economics for the behavior of the marketplace rather than appreciating the vast body of scholarship associated with trying to understand that behavior. This is equivalent to blaming ecologists for such ecological catastrophes as the collapse of the Atlantic cod fishery or the decline of water quality in the Chesapeake. We must distinguish between the systems studied and judgements concerning the state of such systems.

Second, there is a clear need to translate carefully the jargon and vocabulary used within respective fields. In the context of this volume, we must understand clearly what we mean by such terms as "value," "function," and "process." The notion of "marginal value" reflected in the rate at which we are willing to trade off more of one thing for less of something else, is not the same as Whigham's definition of "ecosystem values," the goods and services derived from ecosystems.

Ecology, Economics, and the Cost of Complexity

Simplification, i.e., alteration of ecosystem structure and composition in order to concentrate more of total ecosystem productivity into what humans need or want, has historically been a primary strategy for much of natural resource management. The focus on management, cultivation, or harvest of single species in agriculture, aquaculture, and forestry represents extremes of this strategy. To the extent that alternative management strategies such as riparian corridors, preserves, no-till farming and multiple species management diminish the provision of marketed commodities, their short-term opportunity costs are easily quantified. What is often less clear are the benefits of such strategies.

The first benefit of a retreat from simplification may be insurance against catastrophic risk. Simplification often increases risk of loss at local scales. Extensive simplified ecosystems and landscapes are at greater risk to catastrophic disturbance and are less likely to rebound from such disturbance in a timely manner. Complexity is a hedge against such risk. Thus, the area of risk assessment may provide untapped tools for valuation of ecosystem properties.

The second argument for maintaining complex ecosystems is that simplified ecosystems may make sense at local (and we might also note, private) scales, but taken over entire watersheds such as the Chesapeake, there are cumulative impacts and risks that have significant economic costs. Thus, nutrient subsidization required for single crop farming may be "profitable" at the scale of a 40-acre field, but the cumulative costs to the entire watershed of widespread use of fertilizers may be enormous, and if included in the accounting might redound to an overall loss. In economic parlance, simplification may be privately advantageous, but socially inefficient to the extent that private actions generate uncompensated externalities.

That key ecosystem properties or processes are undervalued in existing markets seems clear. Market forces alone, particularly when focused on short time frames and small spatial scales (i.e., with many externalities) may drive management decisions which in the long term or on large scales are not sustainable. Regrettably, it is much more difficult actually to "get the prices right" (that is, to define appropriate incentives for the maintenance of complex ecosystems) than it is to demonstrate that the unregulated private market economy does not achieve this ideal. In order to create appropriate incentives, we must know by how much the existing prices of ecological assets (including those prices which now are effectively zero) differ from the "right" prices. As several authors noted, the tools provided by such approaches as hedonic price regressions, travel cost studies, and contingent valuation surveys provide at present only very rough guidance.

What if we could get the prices right? If, through collaboration and communication, ecologists and economists were successful in developing models to predict the benefits and costs of specific societal or management actions, would we automatically adopt more sustainable protocols?

In some sense, such models might be viewed as necessary but not sufficient conditions for sustainable management. As is noted in the chapters by Hennessey and Orbach, institutions necessary to reconcile conflicts across scales of space and time are often absent. On this point, however, we might be, if not blindly optimistic, at least cautiously hopeful. Few individuals or cultures have purposefully acted to diminish core processes necessary to sustain functioning ecosystems. As our knowledge base has improved, our perceptions of values have also changed. Moreover, ecosystems have remarkable resilience to recover from insults. As our knowledge of ecosystems increases, we can hope that it will translate into behaviors that will prevent us from crossing critical thresholds that may threaten their sustainability.

Index